Robot Ecology

Robot Ecology

Constraint-Based Design for Long-Duration Autonomy

Magnus Egerstedt

Princeton University Press
Princeton and Oxford

Published by Princeton University Press
41 William Street, Princeton, New Jersey 08540
6 Oxford Street, Woodstock, Oxfordshire OX20 1TR

press.princeton.edu

Library of Congress Control Number: 2021945071

ISBN 9780691211688
ISBN (e-book) 9780691230078

British Library Cataloging-in-Publication Data is available

Editorial: Susannah Shoemaker and Kristen Hop
Production Editorial: Nathan Carr
Jacket/Cover Design: Wanda España
Production: Danielle Amatucci
Publicity: Matthew Taylor and Charlotte Coyne
Copyeditor: Susan Matheson

This book has been composed in Nimbus and Avantgarde

Printed on acid-free paper. ∞

Printed in the United States of America

10 9 8 7 6 5 4 3 2 1

For my favorite traveling companions,
Danielle, Annika, and Olivia.

Contents

Contents

III   Robots in the Wild

Preface

The first seed to this book was planted in Costa Rica when I was there with my family in 2014. During that trip, I got somewhat obsessed with sloths. I could not understand how these slow and potentially quite tasty animals could exist. How come they weren't immediately eaten by eagles or jaguars? What was the point with such slowness? I started reading about sloths and learned about why a slow, low-energy lifestyle is sometimes advantageous. As a result, I got very fascinated with the idea of also embracing slowness as a design paradigm in robotics. Luckily for me, the Georgia Institute of Technology was at that time home to one of the most creative roboticists I know—Prof. Ron Arkin—and he shared my views that this was something worth exploring. We started investigating the role of heterogeneity, broadly interpreted, in multi-robot teams, where slowness (as well as being fast) constitutes one dimension along which heterogeneity can be understood.

Together with Prof. Vijay Kumar at the University of Pennsylvania, Ron and I secured the funding necessary to start investigating the role of heterogeneity in general, and slowness in particular, through Marc Steinberg's Science of Autonomy program at the US Office for Naval Research. Marc is an exciting program manager to work with since he is highly supportive of his investigators taking intellectual risks and going off in sometimes surprising directions. As such, he gave us free hands to explore what we, by now, had started referring to as "robot ecology."

As my foray into the world of really slow robots continued, I realized that I had much to learn about slow animals. And, in particular, I had to strengthen my bond with sloths. But I needed professional help. After a bit of online searching, I discovered the ecologist Prof. Jon Pauli at the University of Wisconsin–Madison, and I decided to send him an email. Sending this email may have been one of the smartest things I've done professionally. As it played such a pivotal role for the developments in this book, it is included verbatim, together with Jon's response.

From: Egerstedt, Magnus
Subject: SlothBots?
Sent: Thursday, February 18, 2016 5:20 PM
To: Pauli, Jonathan

——

Hello Jonathan; I'm a roboticist at Georgia Tech and I have recently become very interested in slowness as a design paradigm - slooooooow robots that "live" for extended periods of time out in the field/farm/forrest. As part of this, I recently got a grant to build SlothBots, i.e., robots that are behaviorally inspired by sloths. I have some funding to connect with biology "consultants" and after a bit of internet searching, it seems like what you do fits perfectly! So, would you be interested in collaborating a bit on my SlothBot project? Maybe come down to Atlanta and give a talk and help us design the slowest robots ever?

Best! / Magnus

From: Pauli, Jonathan
Subject: Re: SlothBots?
Sent: February 21, 2016 at 2:32 PM
To: Egerstedt, Magnus

——

Hi Magnus,
Apologies for the delayed response—last week ended up being unexpectedly hectic. I have to admit that your invitation is possibly the most interesting, and probably coolest, one I've gotten in my career so far. I'd be really interested in visiting with you and your group at Georgia Tech, giving a seminar on our ecological studies of sloths, and chatting about your SlothBot. For full disclosure, though, I am not an anatomist or kinesiologist so its unlikely I'll have academic insights into things like movement mechanics. But, we have been studying sloths in the field (observation, tracking movements, collecting samples for genetic and isotopic analyses as well as studying their demography and determining interspecific interactions). So, if an evolutionary ecologist of sloths works for

you, then I'd be very much game for a visit and to explore this opportunity to collaborate.

All the best,

Jon

Jon and I started collaborating, and as I learned more about ecology and how ecologists think about their beautiful field of study, I started pondering what a formal framework for robot ecology would look like. In what way is the design for robots that are to be deployed over truly long time-scales in natural environments different from the design for robots that perform specific tasks in curated environments, such as in my own academic lab? This is where the second seed to the book comes into play.

In 2016, Prof. Aaron Ames joined Georgia Tech. As our research interests—the intersection between control theory and robotics—were very similar, we decided to hold joint lab meetings to encourage our students to spend time learning from each other. What Aaron and his collaborators had been developing were so-called control barrier functions for encoding constraints. And it did not take long until many of my students talked about control barrier functions in our weekly meetings. As such, Aaron's influence on my lab was immediate and profound. In particular, one of my former students brought barrier functions to robot swarms as a mechanism for avoiding collisions. The idea was that rather than explicitly specifying what the robots should be doing, the barrier constraints would kick in when collisions were imminent and gently guide the robot swarm away from unsafe configurations.

This powerful idea that constraints rather than goal-driven behaviors can give rise to useful, elegant, and complex robotic systems plays a central role in robot ecology. From Jon I had learned that richness of animal behavior can oftentimes be described through the effects of environmental constraints, such as the prevalence (or lack thereof) of food, water, predators, mates, and so on. And from Aaron I learned how to formally think about constraints in a control-theoretic setting. What was missing was the proper way of contextualizing these two ideas in a coherent way, which brings me to the third and final seed to the book.

One afternoon, about a decade ago, I was walking around my lab at Georgia Tech, feeling rather pleased with myself. The lab had become one of the global hubs for swarm robotics research, and we had built up a significant infrastructure to support the research that allowed us to go from idea to full-blown robotic implementation and experimentation without too much of a hassle. But, as I started counting the dollar values associated with all the stuff in the lab, I became increasingly alarmed. It takes a massive amount of resources—equipment, people, space, money—to run a world-class swarm robotics lab. As a result, a number of really great researchers are locked out from participating and are forced to rely solely on simulated robots. I took it upon myself to rectify this inequity problem and in 2015, Dr. Pramod Khargonekar visited the lab. Pramod was at that time serving as the Assistant Director for Engineering at the National Science Foundation (NSF), and I told him that I wanted to build cheap swarm robots for researchers all across the world. He liked my ambition but told me to think bigger. This was the jolt I needed to create the Robotarium.

Funded through NSF's Major Research Instrumentation program, with the enthusiastic support of Dr. Kishan Baheti at NSF, the Robotarium is a remotely accessible swarm robotics lab where researchers, students, and educators from all over the world can upload code, run robotics experiments, and get the scientific data back. After going live in August 2017, many thousands of experiments have been executed on the Robotarium by users from all continents except Antartica. And, most importantly for the robot ecology narrative, the Robotarium has been in continuous operation for many months at a time without the need for any human intervention; it is an absolute joy to see the robots all of a sudden leave their charging stations late at night to go and execute some experiment that is being orchestrated from the other side of the world.

Over Thanksgiving 2019 I came to the realization that not only did these three seeds (slow robots, constraint-based design, long-duration autonomy in the Robotarium) combine together into one coherent story centered on the idea of robot ecology, I also needed to turn the story into a book. Hence the words on this page.

Fast-forward a year, and the final product is organized into three parts, each with its own distinct flavor. Part I is focused on the broad theme of what it means for a robot to be present in a natural environment over truly long time-scales. In particular, the tight coupling between robot and its environment (or, more suggestively, between organism and its habitat) takes us down an ecological path where robot "survival" takes precedence over all other design considerations. Starting off with the inspirational story of the two Mars rovers, *Spirit* and *Opportunity*, Part I covers previous approaches to associate robots and robot behaviors with their deployment environments. It also makes the initial connections to key ecological principles as well as establishes the necessary foundations for how to design useful, scalable, and distributed multi-robot behaviors.

The first part of the book is not overly technical and leaves many questions unanswered. In contrast, Part II is where the formal machinery is introduced that is needed to answer these questions. It shows how control barrier functions constitute the right mathematical objects for reasoning about dynamical constraints which, in turn, leads to a general theory of robot *survivability*. Such a theory relies on the ability to render "safe sets" forward invariant, meaning if the robot starts safe, it stays safe forever (in theory). To arrive at full-blown robot survivability, this construction must be augmented to allow for the production of complex safe sets through Boolean logic, as well as provide support for task persistification. This latter topic is particularly pertinent as it enables robots to be out on deployment over long time-scales—significantly longer than what can be accomplished on a single battery charge.

To put the robot ecology framework on a solid footing, Part II is by necessity somewhat technically involved and requires prior exposure to control theory and dynamical systems to be fully appreciated. Although the book is intended to be approached in a linear fashion, it is perfectly fine to jump directly from Part I to Part III if one were to be so inclined. The third and last part of the book takes the technical developments from Part II and unleashes them on the questions raised in Part I. This is finally where a precise theory of *Robot Ecology* is formulated, and the resulting framework is instantiated on two use-cases.

The *SlothBot*, which is an energy-efficient, wire-traversing, solar-powered, environment-monitoring robot, is designed to be deployed over long periods of time in the tree canopies for the purpose of taking measurements of relevance to the production of microclimate and ecological niche models. As such, it represents a canonical example of the task persistification idea in that its mission cannot be completed on a single battery charge. Similarly, the autonomy-on-demand concept, where robots are present in an environment for long periods of time, tasked with doing mostly nothing at all beyond waiting to be recruited to participate in a wide spectrum of missions, is instantiated on the *Robotarium*. Conceived as a remotely accessible, swarm robotics testbed, the Robotarium has been in (more or less) continuous operation for years, and has participated in thousands of user-submitted experiments. As such, it manifests a number of the central tenets of robot ecology.

A book like this does not appear in a vacuum. Instead, it is the culmination of years of research originating from a number of fruitful and fun collaborations. I have already mentioned the important roles that Ron Arkin, Vijay Kumar, Marc Steinberg, Jon Pauli, Aaron Ames, Pramod Khargonekar, and Kishan Baheti all played in the robot ecology story. I have moreover had the fortune to be inspired, amused, and technically challenged and educated by many other collaborators who have all influenced the direction of this book.

Dr. Emily Coffey, Vice President for Research at the Atlanta Botanical Garden, has taught me a lot about conservation biology, and she has been my co-conspirator when it comes to actually deploying the SlothBot. The Robotarium would not have existed without colleagues at the Georgia Institute of Technology, and Profs. Eric Feron and Raheem Beyah contributed significantly to the initial concept. Additionally, this book contains a number of technical results that were collaboratively discovered. Colleagues whose intellectual fingerprints can be found all over the book include Profs. Sam Coogan (Georgia Tech), Jorge Cortés (University of California, San Diego), Masayuki Fujita (Tokyo Institute of Technology), Calin Belta (Boston University), Daniela Rus (MIT), Evangelos Theodorou (Georgia Tech), Girish Nair (University of Melbourne), Anirban Mazumdar (Georgia Tech), and Seth Hutchinson (Georgia Tech). Even though Seth is listed last, he is the person I have spent the most time bouncing ideas off of, and I cannot overstate his influence on this book (even though he would probably deny that).

I must also give loud shout-outs to Drs. Chris Kroninger and Brian Sadler at the US Army Research Lab who helped focus the autonomy-on-demand concept significantly, and to Susannah Shoemaker—the Mathematics and Engineering Editor at Princeton University Press. When I approached Susannah with the idea of writing this book, she was highly enthusiastic, and there was never any doubt that she would make sure that the book found the right home. It is a lot of fun working with someone who shares your excitement about a project, and she helped me take this from a vague concept to a fully realized book by employing a carefully balanced mix of support and gentle nagging ("Is the book done yet?").

Throughout my career, I have been fortunate to work with a number of talented, passionate, and creative students and postdocs. It is a privilege to be allowed to play a small part in their intellectual journeys and, without a doubt, they are the real forces behind much of what is contained on these pages. The following people were instrumental to the development of the robot ecology framework: Daniel Pickem, Sean Wilson, and Paul Glotfelter took the Robotarium from a vague concept to a fully realized, remotely accessible swarm robotics testbed; Gennaro Notomista and Yousef Emam made the SlothBot into the leisurely-yet-compelling robot it is today; and Li Wang single-handedly made my lab embrace control barrier functions. Additionally, Maria Santos, Ian Buckley, J. P. de la Croix, Yancy Diaz-Mercado, Sung Lee, Pietro Pierpaoli, Siddharth Mayya, Sebastian Ruf, Eric Squires, Mohit Srinivasan, Mark Mote, Anqi Li, Christopher Banks, and Soobum Kim have all contributed to the contents in this book, and hopefully they will find that I have done justice to their elegant discoveries.

I want to end by saying that this book would not exist if it weren't for my fantastic family. The bulk of the writing took place during the COVID-19 outbreak in 2020 as Georgia Tech suspended on-campus activities and we all sheltered at home. When other people took advantage of this new, reclusive lifestyle by baking sourdough bread, making furniture, or learning an instrument, I deprived my family of the prospect of tasty baked goods or a new dining room set. Instead, they got a book about robot sloths. My wife, Danielle, and my daughters, Annika and Olivia, were with me every slow-paced step of the way. Their support, encouragement, and overall enthusiasm for the project are truly what fueled the book. For that I am extremely grateful.

I Long-Duration Autonomy

1 Introduction

Robots are increasingly leaving the confines of their highly structured and carefully curated environments within cages on manufacturing floors, academic laboratories, and purposefully arranged warehouses. This robot relocation is taking the robots to new places, where they are expected to operate across long temporal and spatial scales. For example, in precision agriculture, it is envisioned that robots will be persistently embedded in fields, tending to individual plants by monitoring and meeting their fertilizer, pesticide, or water needs [38, 381]. These agricultural robots will be present in the pastures throughout the full growing cycle, i.e., over an entire season [23]. Similarly, a number of environmental monitoring scenarios have been considered, where robotic sensor nodes are monitoring aspects of a natural environment [124, 392]. Examples include searching for the possibly extinct Ivory-billed Woodpecker in the forests of Louisiana [386], employing underwater robots for tracking marine pollution or the spread of invasive species [189, 407], or for monitoring the effects of climate change on the polar ice caps [388].

1.1 Long-Duration Autonomy

The deployment of robots over truly long time-scales in unstructured environments poses problems that are fundamentally different from those faced by robots deployed in factories or other controlled settings, where operating

3

Figure 1.1: Artist's portrayal of a NASA Mars Exploration Rover [196].

conditions exhibit only limited variability, power is readily available, and regularly scheduled maintenance routines ensure that minor technical problems do not accumulate to produce catastrophic failures. But, in *long-duration autonomy*, robots face a whole new set of challenges [71, 392], and this introductory chapter highlights some of the main themes and opportunities associated with these challenges, as well as makes the initial connection to ecology, i.e., to the *tight coupling between animal (robot) and its habitat (environment)*.

1.1.1   Lessons from Mars

When two Mars Exploration Rovers (MERs), MER-A and MER-B, landed on Mars in January 2004, they were tasked with completing individual missions spanning 90 Martian solar days, which corresponds to roughly 92.5 days on Earth [390]. Better known by their other names, *Spirit* and *Opportunity*, these rovers, as shown in Figure 1.1, managed to outlast their expected life spans by a significant margin and participate in five missions over 6 years and 2 months (*Spirit*) and a staggering 15 years and 1 month (*Opportunity*) [292].

Key to the longevity of these rovers was, of course, a great amount of highly ruggedized hardware and electronics, coupled with carefully designed, stress-tested, and clever engineering solutions. Additionally, the rovers had access to a virtually endless source of solar energy, and their solar arrays could generate as much as 140W per Martian day. Despite the abundance of energy, it was the power system that was expected to be the limiting factor in terms of the duration of the mission, as rechargeable batteries degrade over time and, as such, are no longer able to recharge to full capacity. But the real danger to the power system was the frequent Martian dust storms that not only would block the sunlight, but also accumulate dust on the solar panels, rendering them increasingly ineffective [296].

So why were *Spirit* and *Opportunity* able to perform their tasks significantly longer than expected? The answer was both simple and surprising. The same winds that sometimes caused dust storms on Mars would other times clean the solar panels by sweeping away dust [146]. These co-called "cleaning events" seem to have happened much more frequently than what NASA originally expected. As a result, the solar arrays were kept largely dust-free, and the life spans of the rovers were significantly extended—from less than a year to 15 years, in the case of *Opportunity*.

An immediate lesson one can draw from this interplanetary dust removal anecdote is that interactions between MERs and the environment proved to be beneficial to the rovers. But, at the same time, it was ultimately environmental factors that did the rovers in. *Spirit* got stuck in some particularly soft and sticky Martian soil during the summer of 2009. Despite efforts to free the rover, it was forced to reinvent itself as a stationary "science platform"— a task it performed for almost a year until contact was lost in 2010 [421]. *Opportunity*, on the other hand, did indeed get caught in a massive dust storm during the summer of 2018 that covered the solar panels so completely that it never recovered [421].

By necessity, the rovers were completely reliant on *in situ* solar energy, which, in turn, carried implications for how the robots functioned. One of the more striking manifestations of this dependence on sporadically present sunshine was how slowly the two MERs moved. *Opportunity*, which was the more peripatetic and well-traveled of the two rovers, had completed a full marathon on Mars by March 23, 2015, which translates to a rather leisurely finishing time of around 11 years and 2 months. The reason for this slow and steady pace can be traced back to considerations about energy conservation in conjunction with the need to stay away from trouble at all costs, as it was impossible to rescue a MER after a catastrophic event. As a result, the planning algorithms used for the rovers were highly conservative in terms of uncertainty management [74, 75, 257, 258]. Another way of phrasing this, using terminology borrowed from ecology, is that *survival* took precedent over most other considerations, including any notions of performance-based optimality.

The context in which this book is to be understood is that of long-duration autonomy, and the tale of the two impressive Mars rovers, *Spirit* and *Opportunity*, clearly highlights the two important themes of environmental interactions and survivability.

- Interactions between robot and the environment in which it is deployed play a key role in understanding design for long-duration autonomy; and
- Survivability, i.e., the explicit focus on avoiding getting caught in situations from which the robot cannot recover, takes precedent over all other design considerations.

It should be pointed out that although the MERs were absolute robotic marvels, and significantly advanced our understanding of robotics and autonomy, their operations were not what one would strictly call fully "autonomous." Instead, the rovers employed what NASA dubbed "directed autonomy," where commands were transmitted once per day to the rovers. The commands were encoded as event-driven sequences of motion commands that the rovers parsed using on-board stereo-vision and path-planning algorithms [50]. Despite this technicality, *Spirit* and *Opportunity* provide highly inspirational examples of robots that succeeded at carrying out a series of complex, long-duration missions over truly long time-scales.

1.1.2 Operations Beyond a Single Battery Charge

With the NASA Mars rovers as starting point, and using the key takeaways from their story, we have a handful of promising themes for characterizing and understanding long-duration autonomy. Perhaps the most important (and obvious) observation is that the robots have to be deployed over long periods of time for it to be considered "long-duration." One does not, however, need interplanetary travel to encounter situations where robots may be required to be deployed over long time-scales. In fact, our homes are increasingly being populated by household robots that are more or less in continuous operation, using dedicated charging or waste deposit stations. Environmental robots are being deployed in terrestrial or aquatic ecosystems to monitor factors such as plant growth, pollutants, wildfires, or climate trends, which may require the robots to be deployed for entire seasons. Warehouse robots are expected to perform fetch-and-carry operations; industrial robots are tasked with painting or welding; and mobile guide robots provide information to travelers in airports, art aficionados in museums, or patients in hospitals—all without taking breaks for maintenance or in other ways disrupting operations, e.g., [38, 187, 201, 381].

One way of defining long-duration autonomy is *deployment beyond a single battery charge* (or tank of gas), and where the recharging (or refueling) is part of the robot's portfolio of responsibilities.[1] Note that we phrased this in terms of "deployment" rather than in terms of a long-duration "mission." The reason for this is that we need to allow for situations where the mission may change, or where new missions may be requested. *Spirit* and *Opportunity* were sent to Mars to perform a focused science mission, but as they outlasted their expected life spans, they ended up performing in five different missions with completely different science objectives [292]. Perhaps even more striking and interesting is the situation where the robots may be deployed without any particular mission in mind at all. They are just asked to be present in an environment, waiting to be recruited to do whatever tasks need doing, following an *autonomy-on-demand* model, as opposed to a mission-centric view of what the deployment is supposed to be about [128, 304].

Regardless of whether the deployment involves a single, protracted mission, a sequence of multiple missions, or no clear mission at all,[2] two conditions must be satisfied for it to be considered a long-duration deployment, namely the deployment must last longer than a single battery charge, and the robot must be able to recharge itself.

- *Beyond a Single Battery Charge:* The scope of the deployment must be such that it is impossible for the robot(s) to successfully satisfy the requirements on a single battery charge; and

---

[1]We will use "battery" as shorthand for all sorts of different types of energy sources unless the context requires that the particulars be explicitly called out.

[2]One can of course argue—perhaps even successfully so—that having no mission at all is actually a mission in itself. As we will focus on "deployment" rather than "mission" as the defining characteristic of long-duration autonomy, this conundrum does not really matter for the developments in this book.

- *Autonomous Recharging:* No human intervention can be required in order for the energy sources to be replenished. Instead, the robot(s) must achieve this autonomously.

It is worth pointing out that the first condition, which states that a single battery charge is not sufficient, does not imply that clever power-management is not desired or needed.[3] On the contrary, power-management is certainly playing an exceedingly important role in the successful deployment of robots over long time-scales.

Once the robots are out in an environment for long periods of time, it is quite natural to draw inspiration from other "systems" that are present in environments over long time periods and need to "recharge," namely animals. This connection between animals and their habitats (ecology) and robots and their environments (henceforth known as "robot ecology") is indeed one of the central themes of this book. To this end, a number of biological organisms and habitats will be injected into the narrative in order to highlight and stress particularly salient ecological principles.

1.1.3 On the Value of Slowness

As already hinted at, the impetus behind the NASA Mars rovers' leisurely pace can be traced back to two primary reasons, namely the need to take it slow so as not to jeopardize the robots due to sudden or uncontrolled movements, and the need to conserve energy. As the saying goes, "slow and steady wins the race." Even though it is rare to actually see a tortoise and a hare line up and compete—if they did, the hare would most certainly win—the saying would indicate that the hare also runs a much higher risk of having something unforeseen happen to it due to its hasty outlook. Approaching new situations in hazardous, or even hostile, environments in a careful and deliberate manner is of particular importance when robots are supposed to be deployed over long time-scales, without human intervention. For instance, one of the primary reasons why underwater robotics is so tricky is that it is very hard and costly to recover malfunctioning or lost robots, e.g., [320, 367, 435]. Another manifestation of this idea can be found in the area of *safe learning*, which is

---

[3]Energy (joules) is what is available to the animal/robot, while power (joules per second, or watts) is the rate at which the energy is being delivered as work. So, "energy" will refer to the total charge of the battery, while "power" to the rate at which the battery is being drained.

predicated on the observation that a careless exploration of all state-action pairs can easily lead to the robot finding itself in disagreeable, and even harmful, configurations [5, 28, 45, 426].

Arguably, the primary reason for being slow—among animals as well as robots—is not to be cautious, but to conserve energy. As such, if the available energy is limited, which it usually is in nature, embracing a slow lifestyle can stretch the crucial energy resources further. For instance, *arboreal folivores* inhabit the ecological niche of spending their lives in the trees (arboreal), while sustaining themselves solely on leaves (folivore) [428]. This is a challenging strategy since in order to dwell productively among the trees, animals typically must be small and nimble so as not to simply fall down due to miscalculated leaps or broken branches. Now, contrast this arboreal size constraint with leaf-eating. Leaves are complicated foods in that they can be both toxic and structurally protected. In fact, as plants cannot move around in order to avoid their predators, they must come up with other means of defending themselves, like with thorns or spikes, or by chemical means [207]. Additionally, the cellulose fibers in the plant cell-walls that provide structural scaffolding to the leaves also make them hard to digest. As a result, animals who consume nothing but leaves must have a sufficiently long digestive tract, i.e., have a big enough gut, to break down these complicated foods [428]. The arboreal folivore is thus faced with the opposing requirements of being big enough to break down the food, yet small enough to live among the treetops.

What is the solution to this size dilemma faced by the arboreal folivores? Animals that occupy this ecological niche, such as koalas, two-toed and three-toed sloths, and some lemurs, all have roughly the same size, and they spend the vast majority of their time just sitting there among the treetops, doing nothing other than digesting their food. And when they do move, it is typically happening at an exceedingly slow pace. In other words, slowness has become a response to a severely energy-constrained existence. We will, throughout this book, return to these low-energy lifestyle animals as wellsprings of inspiration. In particular, the three-toed sloth will serve as a particularly suggestive source, culminating in Chapter 8 with the design of the SlothBot, a preview of which is shown in Figure 1.2.

For now, the takeaway from this initial discussion about power-management and slowness is simply that when operating in an environment where unlimited power is not available, and where the deployment specifications require the robot to function beyond a single battery charge, being slow

Figure 1.2: The SlothBot—a slow and energy-aware robot developed to perform environmental monitoring tasks—traverses a cable suspended between trees on Georgia Institute of Technology's campus.

is part of the toolbox. In fact, slowness is one of the design principles that separates long-duration autonomy from its short-duration counterpart.

1.2 Survivability

The shortest path between two points through a space populated by obstacles is obtained by moving as closely to the obstacles as possible [225, 237], as shown in Figure 1.3. Similarly, the fastest way for a car to come to a complete stop at a stop sign is maximal acceleration until the very last moment, and then the driver should slam on the brakes right at the stop sign [230].

Although optimal (minimum distance and minimum time, respectively), both of these strategies are problematic. What if the range-sensors used for detecting obstacles were not properly calibrated? In that case, the robot would hit rather than skirt the obstacles. Or, what if the model of the brake's effect on the car's motion was slightly wrong? In that case, the car might end up coming to a complete stop halfway through the intersection rather than at the stop sign, with potentially lethal consequences. As shown in [120, 208],

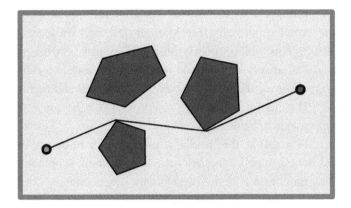

Figure 1.3: The shortest path through an environment typically runs as close as possible to obstacles (the polygons in the figure), thus rendering it non-robust to measurement errors [225, 236].

optimality and fragility are closely related concepts, meaning that optimal solutions are typically non-robust (or fragile) with respect to various types of disturbances, such as measurement or modeling errors. And, in long-duration settings, this lack of robustness can be catastrophic.

1.2.1   Costs and Constraints

Consider the problem of trying to make a solar-powered robot go to a particularly sunny spot to recharge the batteries. There are two different ways of achieving this objective. First, one can define a performance cost that evaluates how well the robot is progressing towards the goal location, e.g., by letting the cost be given by the distance to the goal. We denote this cost by $\mathcal{C}_{charge}$, and the controller should be chosen such that $\mathcal{C}_{charge}$ is minimized, either by incrementally moving against the gradient of the cost [251], or by finding the overall best strategy that minimizes the cost, e.g., by employing some path-planning method [225].

The second approach would instead be to define a constraint, abstractly encoded as $x \in \mathcal{G}_{charge}$, where x is the state of the robot, and where the constraint could say something like "the robot should always be able to make it back to the charging station given the current energy levels." The robot would then be allowed to move freely as long as it did not violate the constraint.

Additionally, the purpose of deploying this robot is likely more than simply making it go and recharge. Instead, the robot is probably expected to perform some primary task, such as detecting interesting events, protecting an area from intruders, or collecting Martian dust, and we let $\mathscr{C}_{task}$ be the cost that encodes how well (or poorly) this primary task is being performed. The question then arises: How should one balance these two different and potentially opposing requirements of recharging batteries and performing the primary task? The answer to this question depends on the context in which the overall mission is to be understood.

Consider, for example, the peculiar lifestyle exhibited by mayflies. Once they are past their nymph stage, they have an extremely short life span—as short as a few minutes for the female *Dolania americana*—and their single focus as adults is reproduction [186, 223]. This strategy leaves them with no need to feed and, as a result, they do not even have fully functioning mouthparts. Classifying this as long-duration autonomy would be a stretch. Similarly, if the robot, just like the mayflies, is supposed to execute the primary task and then be done, and the primary task can be completed with a single battery charge, then one should just ignore the charging requirement and solely minimize $\mathscr{C}_{task}$. One could call this approach the "short-duration autonomy" approach, and it is how robotics algorithms are typically approached. But, as already discussed, a key attribute of long-duration autonomy is that the robot should avoid catastrophic failures at nearly all costs since, once it has failed, there is no recovery. And, getting stuck somewhere in a dark corner of the deployment domain with completely depleted batteries certainly counts as a mission-ending failure. Borrowing, once again, from ecology, this translates to ensuring the survival of the robots.

With the notion of survival added to the mix, one approach could be to impose scheduled behaviors, e.g., periodic visits to the charging stations, during the execution of the primary task. The robot would thus switch between minimizing $\mathscr{C}_{task}$ and $\mathscr{C}_{charge}$ in response to a power-management scheme, e.g., [204, 374, 395]. But adhering to such fixed policies during execution without regard to implications on achieving task goals is not ideal as the robot would either solve the task it is supposed to solve, or recharge. But why not do both?

An attempt at doing both would be to encode survivability as a performance objective and somehow combine it with the other performance goals. This could, for example, be done via scalarization, where the overall performance cost would be given by a combination of the two costs, $\sigma\mathscr{C}_{task} + (1-\sigma)\mathscr{C}_{charge}$, for some $\sigma \in [0, 1]$. Alternatively, a so-called multi-objective optimization approach could be used. In the former case, the primacy of survivability is not ensured unless $\sigma = 0$, raising the possibility of catastrophic failures in the opportunistic pursuit of short-term gain. In the latter case, none of the available equilibrium or optimality concepts (e.g., Nash or Stackelberg equilibria, or Pareto optimality) ensure survival, instead balancing, in one way or another, the degree of survivability against other task performance criteria. In short, these types of approaches fail to adequately recognize (or exploit) the fact that *surviving is a prerequisite to thriving in long-duration autonomy applications.*

As will be seen in subsequent chapters, it is indeed possible to focus on the primary objective, yet ensure the survival of the robot. Constrained optimization provides the appropriate semantics for describing such an outcome. In other words, recognize the primary optimization or optimal control problem, but add in constraints that ensure survival by letting the robot solve,

$$
\begin{aligned}
&\text{minimize } \mathscr{C}_{task}\\
&\text{subject to } x \in \mathscr{G}_{survive},
\end{aligned}
\tag{1.1}
$$

where $\mathscr{G}_{survive}$ could be equal to $\mathscr{G}_{charge}$, or it could be a more general constraint that contains a number of other survival considerations as well, such as avoiding collisions or staying connected to other robots [128]. This seems like a highly promising way of abstractly capturing what long-duration autonomy could be about.

1.2.2   Robots that Do (Almost) Nothing

A particularly pertinent choice of performance cost, $\mathscr{C}_{task}$, in the previous section is to measure how much energy the robot is expending. In the absence of additional constraints, the optimal strategy would thus be to simply let the robot do nothing, i.e., to let the actuators exert no forces or torques on the system, which takes us close to the strategy employed by the arboreal folivores during long stretches of their existence. In fact, the conservation of energy is central to virtually all living organisms and, according to [357], the "purposeful expenditure of energy" is one key characteristic of what it means to be alive. As such, an initial, biologically motivated (yet mathematically vague and, for now, potentially ill-posed) attempt at formulating a design principle for robot ecology would be to modify the constrained optimization problem in Equation 1.1 to the following optimization problem,

$$
\begin{aligned}
&\text{do as little as possible}\\
&\text{subject to } x \in \mathscr{G}_{survive}.
\end{aligned}
\tag{1.2}
$$

One way of interpreting this formulation is as extreme, existential nihilism—
the meaning of life is to expend as little energy as possible, while barely
subsisting. Although a bit depressing, it is entirely consistent with basic
ecological principles, where an animal's behavior is understood in large part
through an energy-balance calculus [357].

Imagine now that a team of robots has been deployed in an environment,
and are prepared to perform whatever tasks might be asked of them. One
could call this setup *autonomy-on-demand*, and between tasks, the robots
should merely be present in the environment, not running out of energy,
and mostly do nothing. In such a scenario, the sloth is a good role model,
and conducting oneself according to the constrained optimization problem in
Equation 1.2 seems like a reasonable strategy. Once the robots are recruited
to perform some task, the corresponding performance cost, $\mathscr{C}_{task}$, can be introduced, and the robot switches from the problem in Equation 1.2 to that in
Equation 1.1.

But, beyond doing nothing, what the autonomy-on-demand framework
suggests is the possibility of having robots with free time. The question then
becomes, what should these robots spend their time doing? There are indeed
opportunities afforded by being a robot of leisure. It could, for example, improve its skills by learning and exploring better control policies. It could also
learn completely new skills. This is a bit more delicate as the robot is literally
tasked with doing nothing, and most of the machine learning apparatus requires some sort of goal or reward against which the suitability of the control
actions can be evaluated. In the absence of such goals, one instead needs to
move towards a more "curiosity-driven" learning paradigm [217, 393, 416],
where the robot explores state-action pairs without a predefined, clear goal, or
where mismatches between actual and modeled effects are being pursued for
the purpose of getting more accurate models of the robot's capabilities. But,
perhaps most importantly, interactions between environment and robot can be
better understood.

The interactions between robot and its habitat is imperative to the robot
ecology framework in that this coupling must be understood and leveraged in
order for the robot to successfully dwell in an environment over sustained periods of time. This is the topic of the next section, and an example is shown in
Figure 1.4, where a robot, using computer vision, must learn to discriminate
between objects according to their texture and color profiles. For instance,
the robot should learn to tell tall grass from boulders, as its ability to traverse these "objects" is completely different. And, the only way to gauge the

Figure 1.4: Based on an object's texture and color properties, a robot must learn which objects in the environment can be traversed and which cannot.

"traversability" of a particular type of object is to interact with it, e.g., to try to drive through it without getting stuck [396].

1.3 Coupling Between Environment and Robot

As already discussed, survivability, i.e., the ability to avoid situations where survival can no longer be ensured, can be naturally encoded as a constraint rather than as a performance objective. This way of formulating survivability is also consistent with ecological principles, where richness of behavior is a direct function of environmental constraints [308, 357, 385], including the abundance and distribution of resources, favorable microclimates, and the prevalence of suitable mates or predators. Indeed, when ecologists study the distribution of species and the composition of populations and communities, the environmental reality and the associated ecological constraints are as important, if not more so, than any "goal-driven" behaviors [357, 385].

Based on this observation that constraints are fundamentally important to animal behavior, one can thus ask if the constraint-based vantage point translates to effective control design principles for engineered systems as well. As such, we will approach the design problem as one where the robots' behaviors are mostly constraint-driven, such as avoiding collisions with obstacles or other robots, or never completely depleting the batteries, as opposed to goal-driven. In fact, these types of constraints can be derived (albeit subject to a slight robotic reinterpretation) from basic ecological principles. As ecology is aimed at understanding the interaction of organisms with their environments and with other organisms, this is a particularly fruitful metaphor also for robots leaving the highly curated laboratory or factory settings, and entering dynamic, unstructured, natural environments across long temporal and spatial scales.

1.3.1 Ecosystems

As was discovered by the Mars rover team, the connection between robot and environment was even more important for the longevity of the robots than what was originally thought. Not only was the environment a source of energy, it was an existential threat through dust build-up and soft and sticky sand. But it also provided unexpected help when the Martian winds would swipe the solar panels clean, thereby overcoming other, more adverse environmental factors. What this anecdote tells us is that the robot and the environment it inhabits should be thought of as a single system, which brings us within striking distance of the idea of an ecosystem.

In the 1930s, a vibrant discussion took place among ecologists about the proper way of thinking about these interconnections, and the term "ecosystem" was coined by the British ecologist A. G. Tansley in 1935. He writes [402]: "The more fundamental conception is, as it seems to me, the whole system (in the sense of physics), including not only the organism-complex, but also the whole complex of physical factors forming what we call the environment of the biome—the habitat factors in the widest sense. *Though* the organisms may claim our primary interest, when we are trying to think fundamentally we cannot separate them from their special environment, with which they form one physical system." This way of thinking about organism (animal/robot) and environment as part of the same system—not in a loose, metaphorical sense but in a tight, physical sense—will prove to be a fruitful way of approaching long-duration deployments. In fact, as animals and plants live (literally) in a physical environment, their *form* and *function* must obey the rules of the physical world [385].

An illustrative example of how form is determined by the physical environment is the size of the pores in avian eggshells. As gas is constantly passing through the eggshell throughout the incubation period to deliver oxygen and nutrients, the movement of the gas follows a diffusion process, which means that environmental factors such as altitude, temperature, and humidity all matter to the type of egg (size of pores) the bird lays [357]. Similarly, rates of processes (r) and animal dimensions (d) typically satisfy an "allometric" relationship, $r = ad^b$, with b being the *allometric constant* [86]. For instance, heart rate versus body mass has an allometric constant of $b \approx -0.2$ among mammals, while the metabolic rate vs. body mass has $b \approx 0.7$ [86, 357]. In other words, chipmunks have a higher heart rate but a lower metabolic rate than elephants. And these environmentally informed form factors have implications for the animals' functions.

On the functional side, animals constantly move among so-called environment patches as the environment changes, over days, months, and even years [357, 385]. And, for the purpose of this book, functional considerations will play a more prominent role than form considerations. That is not to say that form does not matter—it does. Only that the focus of this book is on the control design considerations when deploying robots over long time-scales, i.e., hardware will play second fiddle to software.

There are a number of situations where this idea of functional coupling between robot and environment is not only useful, but crucial when deploying robots over long time-scales. The most apparent and covered situation is the recharging of batteries using energy from the sun, meaning that the robot must, every so often, find itself in a place with ample sunlight. In other words, cave-dwelling robots must either surface every now and then to bask in the sun, or they must rely on some other source of energy. However, energy harvesting does not provide the only beneficial coupling between robot and environment. When aerial gliders or marine robots move through their domains, updrafts and ocean currents, respectively, provide opportunistic sources of low-cost mobility [6, 147, 373]. And passive walkers, e.g., [432], only really function in worlds consisting solely of gentle downhill slopes—in all directions.

1.3.2 Natural and Engineered Environments

So far, the discussion has been focused on natural environments. But, the importance of harnessing the coupling between robot and its habitat is certainly not diminished in engineered environments. Robots roaming around in warehouses or homes must not only be able to locate outlets, they must be equipped with the proper hardware (e.g., plugs) to allow them to take advantage of the available energy sources.

From a mobility perspective, environment and hardware design go hand-in-hand, and one reason why humanoid robots are deemed particularly useful as companion robots in our homes is that our domestic environments are already built for bipedal humanoids that are four- to six-feet tall, with staircases that reward legged locomotion and doorknobs that are strategically placed at certain heights. In other words, our homes are already superbly well-suited for humanoid robots [156]. One does not, however, have to look to the world of autonomous robots to see this phenomenon. For more familiar and mundane vehicles, a carefully engineered environment is oftentimes called for. Trains are extraordinarily well-positioned to take advantage of train tracks, airplanes of airport runways, and cars of highways. We even sometimes modify animals to make them fit our engineered environments, such as putting shoes on horses or electric collars on dogs.

As we will see in a later chapter, this idea of making slight modifications to the environment for the explicit purpose of rendering the robots' existences more productive and safe will prove beneficial when deploying robots up in the treetops in persistent environmental monitoring applications. Climbing, as practiced by arboreal animals, is problematic from a safety point of view in that a robot that falls out of a tree will probably not be able to continue on with its mission. But, by stringing cables in the treetops, we can ensure that the robots can dwell successfully in the tree canopies and remain safely suspended, even when actuators fail or energy levels drop precipitously.

Regardless of whether the robots are to be deployed in jungles, on train-tracks, or in kitchens, when the deployments transpire over long temporal scales, unexpected things are inevitable [71]. The world is fundamentally a messy place, and any attempt at enumerating all the possible things a robot might encounter, in all but the most sterile environments, is doomed to fail. As such, the strategies employed in long-duration settings must support adaptation to new situations. And they must achieve this while ensuring the robot's safety at all times and at (almost) all costs.

1.4 Summarizing and Looking Ahead

What this introductory chapter has done is paint a mood picture and describe some of the challenges associated with long-duration autonomy. It also identified a collection of guiding principles that permeate the book and that illustrate why long-duration autonomy, as compared to its short-duration counterpart, is different from a control design vantage point.

These principles combine together under the umbrella of *Robot Ecology* as follows:

- The tight coupling between robot and environment is not only important, it is absolutely crucial if the robots are to exhibit longevity;
- Survival is a prerequisite to thriving, i.e., rather than minimizing performance-based costs, the control design should focus on ensuring that various safety constraints are satisfied;
- Key among safety constraints is power-management since, if the robot finds itself with depleted batteries without any ability to recharge, it is game over;
- When the deployment takes place over truly long time periods, being fast is oftentimes both energetically wasteful and dangerous, and the robots should embrace a slow lifestyle; and
- Any attempt at enumerating everything the robot may encounter is doomed to fail, and the ability to adapt to changing environmental conditions and missions is a core attribute in long-duration autonomy.

The remainder of this book will take these casual observations and make them more precise and mathematically well-defined. Part II of the book will serve this purpose by establishing control barrier functions (CBFs) as the proper framework for talking about robot survival using the constrained optimization semantics. Once CBFs have been adequately introduced, they will be drawn on to support persistified robot tasks, i.e., to modify nominal controllers in order to extend the robots' life spans indefinitely. The developed tools and techniques will then be employed in Part III for long-duration deployments in a number of different settings, with a particular focus on environmental monitoring and conservation tasks. However, before these tools can be unleashed, a more thorough discussion is needed of what robot survival actually entails, and of how ecological principles can be put to use towards the overarching theme of establishing a theory of robot ecology.

2 Survival of the Robots

Once the shift is made from a goal-driven, performance-centric design outlook to one where survivability rules supreme, one needs to untangle what this actually means technically. As already discussed, constraints constitute prime candidates for encoding survival under the banner of robot ecology. However, there are a number of other potential ways in which this can be (and has been) approached. This chapter sets the stage for subsequent developments by connecting to contrasting ideas for ensuring robot safety, such as using dedicated, prioritized safety "behaviors."

It is worth noting that *survival*, and its close cousin *safety*, only become meaningful constructs when they are contextualized, since insisting on robots not crashing into obstacles is rather pointless (and trivial) in a world without obstacles. In fact, the already introduced, tight coupling between robot and its habitat (deployment environment) asserts itself once again as critical when developing a theory of long-duration autonomy. Of course, the idea that robot design should be understood and control strategies constructed through the lens of robot-environment interactions is not new. In the monograph *Behavior-Based Robotics* [25], Ron Arkin states that

Robotic behaviors should match their environment well, that is, fit a particular ecological niche.

This certainly sounds like a statement very much in line with the robot ecology theme. Although the mathematical formalism pursued in this book will be quite different from previous attempts at characterizing these environmental couplings, we, in this chapter, investigate what it means for a robot to *survive* over long time-scales by situating the long-duration autonomy concept relative to other robot architectures.

2.1 Behavior-Based Robotics

As already observed in the introductory chapter, one attempt at ensuring that a robot is safe is to switch between different controllers, thereby making sure that safety controllers are given ultimate authority over the robot if need be. Although a different vantage point was presented using the semantics of constrained optimization, prioritization seems like a perfectly reasonable attempt at approaching the safety issue.

Following the definition in [25], a *behavior* is a "stimulus/response pair for a given environmental setting." And, if the environment is such that the robot is driving rapidly towards an obstacle, the response from the obstacle-avoidance behavior should be to slam on the brakes or steer away from the obstacle, or both at the same time. If the so-called *arbitration mechanism*, i.e., the mechanism whereby control authority is given to different behaviors, is such that the collision-avoidance behavior is solely in control, the obstacle is guaranteed to be avoided (provided that the avoidance behavior is properly designed), and the robot gets to live to fight another day. This is an example of a *behavior-based* control architecture, e.g., [20, 22, 25, 62], where different behaviors (or control strategies) are tailored to achieve different objectives, and some prioritization scheme is employed to select the active behavior(s) among the set of all available behaviors. This simple yet powerful idea is illustrated in Figure 2.1.

For such a priority-based strategy to be successful, e.g., for it to be able to ensure that the robot is indeed always safe in the sense that it not only tries to do the right thing, but it also reacts quickly enough to the sensory inputs, the response must be fast enough given the current state of the robot and the environment. This idea of closely coupling the robot's response to direct environmental "stimuli" has a rich history, and one can produce remarkably intricate and sophisticated robotic responses using simple rules. W. Grey Walter proposed such a design already in the mid-1950s, named *Machina speculatrix* [422], that was later realized in hardware under the moniker

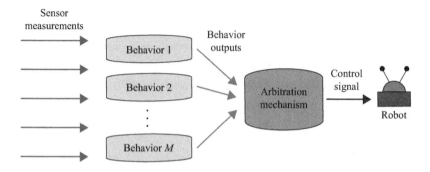

Figure 2.1: An arbitration mechanism selects from a set of possible behaviors as a way of establishing robot safety in the face of unknown and dynamic environmental conditions.

"Grey Walter's Tortoise" [188]. What this robot did was follow a series of *prioritized behaviors*, including seeking out light sources, moving towards said light sources, and avoiding obstacles. Here, avoiding obstacles would be given higher priority than moving towards light sources, which, in turn, took precedence over the light-seeking behavior. The result from this experiment was quite suggestive, and the robot would move around in what seemed like a highly purposeful exploration of its environment, searching for light sources.[1]

Although no formal guarantees were given for the overall performance of Walter's system, it sure did point towards the idea of systems that would perform some primary task (like exploration) and every now and then would go and recharge the batteries based on the discovered resources (light). In fact, one could argue that the developments in Chapter 5 of this book, focusing on how to render primary robotic tasks *persistent*, is taking the *Machina speculatrix* idea and putting it on firm, mathematical footing. In that chapter, robots are tasked with exploring an environment for the purpose of detecting events and environmental phenomena of interest, all the while ensuring that the battery levels never get depleted by, every now and then, taking detours to sunny spots to recharge the solar-powered batteries.

2.1.1 Behaviors in Robots and Animals

At the center of behavior-based robotics is, of course, the notion of a "behavior." Unpacking the suggestion that a behavior is a response to a given

---

[1]Another early instantiation of this idea, where a number of so-called inhibitory and excitatory behaviors were used to generate a rich class of robot motions, is given in [59]. These rules created a varied and compelling set of so-called Breitenberg's Vehicles.

environmental condition, this would abstractly translate to a map, $\beta : X \rightarrow U$, where X is the space where not only the state of the robot takes on values, but also where environmental states are contained, and U is the set of possible control actions, as interpreted broadly. Typically, the set U is not overly problematic in that the robot dynamics, i.e., the mathematical characterization of how the robot moves/evolves, is given by a controlled differential equation, where the control term $u \in U$ is explicitly included. The set X, however, is a bit more involved.

The robot model will inevitably contain a notion of state, x, which will commonly live in some vector space, e.g., $x \in \mathbb{R}^n$, describing things like the robot's position, orientation, and translational and rotational velocities. Yet, we also need to capture environmental conditions as well, since the interactions between robot and environment is of paramount importance [40, 114, 167, 224, 346, 405]. Factors such as the locations of obstacles or other robots, the prevalence of sunlight (which, by the way, is time varying), or the degree to which a particular terrain is navigable, all play their parts when describing the way the robot can interact with the world. For now, we will punt until Chapters 4 and 5 to address the technical composition of X, and will simply assume that we understand, at least at a high-level, what is meant by the joint environment-robot state space, X.

One of the recurring themes in behavior-based robotics is that the robot should act on the highest priority behavior applicable in a given situation for the purpose of the safety and survival of the robot, which is ultimately why behavior-based robotics is a topic of significant relevance to robot survival. But, there are naturally other possible behaviors worth pursuing beyond safety behaviors. In [24, 25], a number of such behaviors is given, organized around the way in which they can be leveraged to interact productively with the environment.

Behavior Classification	Example Behaviors
Exploration	Wandering, covering an area, searching
Goal Seeking	Seeking an object or area
Safety	Avoiding collisions, protecting a boundary
Navigation	Following a road or a stripe
Cooperation	Foraging, flocking, forming a shape, approaching the same location

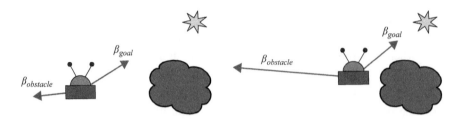

Figure 2.2: As the robot gets closer to the obstacle, the magnitude of the obstacle-avoidance behavior vector increases, thereby indicating that this behavior has a higher priority.

Some of these behaviors constitute positive "tropisms" in that they attract the robot towards a particular object/region/direction, such as when "seeking an object" or "following a road," while others are negative tropisms, such as when "avoiding collisions." The way these attractions and aversions (also known as excitatory and inhibitory influences) are encoded is not *a priori* given, and a number of different such representations have been proposed.

To provide an example of how behaviors can be expressed, consider a planar, kinematic robot, whose state is given by its position and orientation. Using the "motor schema" formalism put forth in [21], one can let a behavior be given by a two-dimensional vector, whose direction is dictating the direction in which the behavior guides the robot to move, and its magnitude is weighted by the priority of the behavior. The way one would approach robot survival in this context is by increasing the magnitude of a given behavior as it becomes more critical, e.g., avoiding obstacles should matter more when closer to obstacles than when farther away. Once an obstacle is close enough, it should be the only thing that really matters to the robot, as illustrated in Figure 2.2.[2]

As reaction times (relative to the time-scales of the environment) and the resulting "urgency" of the behaviors are of the essence when trying to ensure robot survival, the distinction is sometimes made between "reactive" and "deliberative" behaviors, where the reactive behaviors are more immediate, e.g., purely algebraic mappings from perception to action, without the use

---

[2]Even if one has committed to encoding behaviors as vectors whose magnitudes correspond to their priorities, it is not evident how the behaviors should be combined. One could add them together using vector superposition, where the behaviors with higher priorities matter more than others. But one could also envision a winner-takes-all kind of arrangement. This behavior "combination" is referred to as an arbitration mechanism, and it is the topic of a subsequent section.

of intermediary world-representations, such as maps or object classifiers, in order to produce "timely robotic responses in dynamic and unstructured worlds," to borrow a quote from [25]. This tight coupling obviously provides a nod towards survivability as it places the immediate nature of the action front and center. This is also the case in nature, where reflexes and even dedicated, special purpose neural systems have been observed to produce immediate responses to stimuli, like cockroach locomotion, visuomotor control and coordination in flies, and even scratch reflexes in turtles, e.g., [20, 21, 22].

And, speaking of which, the animal kingdom provides a wealth of opportunities when it comes to behaviors. Robotics has, repeatedly, gone back to this well for the purpose of both bio-inspiration and bio-mimicry,[3] e.g., [16, 70, 115, 190]. *Ethology* is the study of animal behaviors under natural conditions. In that discipline, it is stressed that behavioral studies must primarily take place in the wold, as an animal's response can only be meaningfully interpreted in its natural setting [248], which certainly resonates with the ecological idea of a complete system, where the animal (and its behaviors) is but one constituent part of the ecosystem in which it resides. As such, this seems like a promising place to originate a theory of survivability for long-duration deployments.

Survival is not fully a matter of always being fast, as shown with painstakingly slow-paced clarity by the Mars rovers. In fact, it comes down to *timely* responses to external stimuli, where some have to be fast, like moving away from an immediate danger, while others should be slow, like migrating to a more amiable territory. As such, one ethological way of classifying animal behaviors is based on their time-response profiles, as is done in [41, 270], with the classes being (from fast to slow):

- *Reflexes:* Rapid, involuntary responses triggered by environmental stimuli;
- *Taxes:* Responses that direct the animal toward or away from stimuli; and
- *Fixed-Action Patterns:* Responses that persist for longer than the stimuli themselves.

---

[3]The difference between bio-inspiration and bio-mimicry is that the former concerns itself with finding solutions to problems that are inspired (sometimes rather loosely) on arrangements encountered in the natural world. The latter, on the other hand, has the goal of precisely replicating biological designs. Bio-inspired researchers ultimately do not necessarily care if they get the biology right. Bio-mimicry researchers most certainly do.

A hierarchical structure that has a similar organization around how "urgent" an animal's behavior might be, from a survival vantage point, is provided in [248], with urgency being organized from fastest to slowest,

- *Fast:* Chasing, fighting, displaying
- *Medium-Fast:* Parenting, courting, nesting
- *Medium-Slow:* Guarding territory
- *Slow:* Migrating

This way of thinking about animal behaviors in terms of their sense of urgency and time response profiles naturally foreshadows a formal notion of survivability, as will be seen in subsequent chapters, where a constraint-based formalism allows for different survival concerns to evolve according to their own, natural time constants, as encoded through the system dynamics. *The main conclusion to draw from this discussion about behavior-based robotics is that the organization of robot actions as more or less immediate responses to sensory stimuli allows for a design where more pressing safety and survival concerns are given precedence over less urgent activities. Fighting off predators takes precedence over migration (animals), or not driving into things takes precedence over finding lost objects in the environment (robots).*

2.1.2   Arbitration Mechanisms

Given a robotic system that has been designed around a number of behaviors, dedicated to performing targeted tasks or mapping external stimuli to motor control actions, the next order of business is to decide which behavior to use in which situation. And, viewed from afar, a basic organizing principle in behavior-based robotics is to always pay sufficient attention to, and subsequently act on, the highest priority behavior in a given situation [25, 62]. But there are a number of different ways in which this organizing principle can be interpreted, e.g., [20, 22, 59, 62, 64, 129].

What must be decided is thus, when a robot is equipped with a collection of relevant behaviors, how should it choose which behavior (or which set of behaviors) to execute given the particular environmental circumstance it finds itself in? Additionally, if more than one behavior is allowed to be active and influence the motion of the robot at any given time, the combination of these behaviors must be properly framed. Returning to the situation of a planar robot, with position $x \in \mathbb{R}^2$, and adhering to the program of letting behaviors be encoded by a vector $\beta_i(x, x_e) \in \mathbb{R}^2$, where i is the index of the behavior and x_e is the environmental state—as perceived by the robot—the motor schema interpretation associates the priority of Behavior i with the magnitude, i.e., priority(β_i) = $\|\beta_i\|$.

If Behavior i is the only relevant/active behavior in a given environmental circumstance, and if we let β encode the output from the arbitration mechanism, we would have the trivial output $\beta = \beta_i$, in this particular case. Correspondingly, one should move the robot in the direction dictated by the behavior, e.g.,

$$\dot{x} = \kappa \frac{\beta}{\|\beta\|}. \tag{2.1}$$

Here, $\beta/\|\beta\|$ is a unit vector pointing in the direction of β (the direction the behavior is telling the robot to move in), and κ is a positive gain—possibly state-dependent—that corresponds to the speed at which the robot is prescribed to move.[4]

But what should happen when more than one behavior is claiming to be relevant in a particular situation? One choice of arbitration mechanism could be to use a winner-takes-all strategy, whereby the highest priority behavior completely dictates what the robot should do (as was the case for Grey Walter's tortoise [422]). If we continue the convention of letting β be the output of the arbitration mechanism, and let $\mathscr{A}$ be the set of active behaviors, the winner-takes-all approach is given by

$$\beta = \beta_{i^\star}, \quad \text{with } i^\star = \operatorname{argmax}_{i \in \mathscr{A}} \{\text{priority}(\beta_i)\}. \tag{2.2}$$

This strategy is shown in Figure 2.3 (Top), where $\mathscr{A} = \{1, 2, 3\}$, and where priority($\beta_2$) > priority($\beta_1$) and priority($\beta_1$) > priority($\beta_3$). As a result, $\beta = \beta_2$.

Using the terminology from the previous chapter, the winner-takes-all strategy corresponds to hard switches among behaviors. Although appealing in that survival might be ensured by letting the most pressing concerns be addressed completely, this way of switching among behaviors is not unproblematic from a robot ecology vantage point. For instance, how do we know that the prioritization is done correctly? What if a robot is both about to run out of battery charge and slam into an obstacle? Should it go towards the charging station or avoid the obstacle? Which one has a higher priority?

---

[4]This controller is just one of many different possible controllers that move the robot in the direction dictated by β. It is to be thought of as nothing more than an illustrative example rather than a definitive statement about robot control.

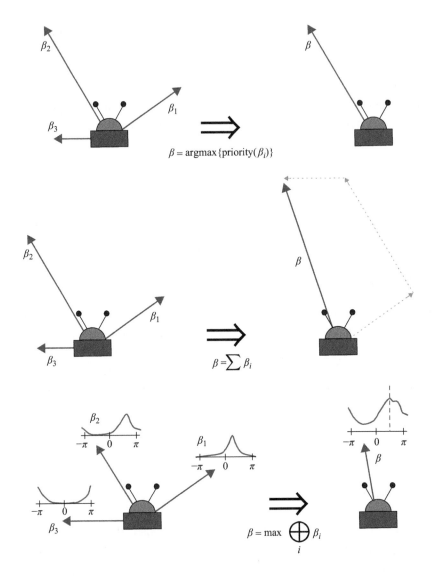

Figure 2.3: Arbitration mechanisms. Top: The behavior with the highest priority (largest magnitude) is selected based on a winner-takes-all strategy. Middle: A blended strategy is employed, where the behaviors are combined using vector superposition. Bottom: Each behavior provides a vote over the possible robot headings and the final behavior is steering the robot in the direction that receives the highest, combined vote, abstractly denoted by the $\oplus$ operator in the figure.

If the obstacle is avoided without consideration for the location of the charging station, the robot may remain dent-free yet end up stranded with completely depleted batteries without any means for recharging. Ever. Or, if it prioritizes going to the charging station, it may break by running into the obstacle. The solution, of course, would be to somehow do both at the same time.

In the original motor schema formulation [20], behaviors were combined, or "blended," based on their priorities using a direct superposition of the behaviors,

$$\beta = \sum_{i \in \mathcal{A}} \beta_i. \tag{2.3}$$

Here, the priorities are implicitly taken into account as more urgent behaviors have a larger magnitude and, as a result, contribute more to the vector summation than smaller-magnitude vectors. Although the resulting vector, β, may have a significantly larger magnitude than the constituent behaviors, this is not a problem as the normalization in Equation 2.1 ensures that only the direction of β matters. The effect of using vector summation as the arbitration mechanism is shown in Figure 2.3 (Middle), where Behavior 2 still matters more than the other behaviors by virtue of its higher priority. From a survival vantage point, this way of doing things seems reasonable—at least most of the time. However, it is also clear that there are situations where a winner-takes-all strategy is the only feasible way to go. If the robot is about to experience a catastrophic failure unless a single, critical act is performed, blending behaviors does not seem prudent.

So what is an arbiter to do if the winner-takes-all strategy is too focused on achieving single tasks at the expense of the overall mission, while a blended strategy can miss the mark in terms of ensuring robot survival? A slight twist on the theme of vector summation that allows for quite a bit of generality was provided by Julio Rosenblatt's DAMN (Distributed Architecture for Mobile Navigation) framework [361]. The idea is to let the different behaviors vote over the set of possible actions. In the case of the planar robot, where the behaviors provide desired headings for the robot to steer towards, the votes could for instance be a function $\beta_i : [-\pi, \pi] \rightarrow [0, 1]$, where the behavior is no longer a vector but a function that associates a vote between 0 and 1 to each possible heading in which the robot could move in. By adding all the behaviors together, as is done in Figure 2.3 (Bottom), one would get a new function defined over the domain $[-\pi, \pi]$, and the direction of travel would simply be the heading that receives the highest overall vote.

Figure 2.4: Three different arbitration mechanisms. Left: The robot is blending the effects of a go-to-goal and an obstacle-avoidance behavior. Middle: By switching between the two behaviors, a winner-takes-all strategy is employed. Right: A constraint-based strategy is used to make sure that the robot navigates towards the goal while staying away from the obstacle.

Variations to this theme have been envisioned, where, for example, negative votes are allowed. Or where vetoes (do not, under any circumstance, move in a vetoed direction), or even impulse votes (move in a particular direction no matter what the other behaviors might be saying) can be used to enforce certain behaviors [65, 334, 434]. From a survival point of view these more general interpretations of the voting rules take us closer to something that resembles constraint-based control design in the sense that hard constraints can be enforced, e.g., through vetoes, so that the robot moves as close as possible to some desired heading (as specified by some performance-based behavior), while staying away from unsafe regions (as specified by the safety behaviors) no matter what.

In Figure 2.4, these three different forms of arbitration mechanisms are implemented on a mobile robot, based on [129]. In the left figure, a blended (vector superposition) strategy is used to produce a smooth but not necessarily safe robot motion by combining a go-to-goal behavior with an obstacle-avoidance behavior. In contrast, the middle figure depicts what happens when a winner-takes-all strategy is utilized to have the robot switch between two behaviors. As can be expected, the result is somewhat abrupt and the robot ends up switching quite a few times between the two behaviors. The right figure shows what happens when the go-to-goal behavior votes for desired directions of travel but the obstacle-avoidance behavior acts as a constraint by vetoing certain directions of travel that would otherwise lead to a collision. From a long-duration autonomy and robot ecology perspective, this third option seems most promising in that the result is both guaranteed to be safe, while progressing the robot towards the completion of a task, such as reaching a goal location. In fact, this approach will be reinterpreted in a control-theoretic setting in Chapter 4 as a way of achieving provably safe yet performance-centric robot motions.

2.2 Multi-Robot Behaviors

So far, safety has primarily been understood and discussed in terms of not colliding with obstacles. But, if more than one robot is participating in the long-duration mission, not colliding with fellow robots matters as well. And it is important to note that other robots are not just moving obstacles. They can be collaborative and friendly, indifferent, or even malicious, e.g., [142, 276]. To this end, the realm of multi-agent robotics and multi-robot behaviors requires its own, dedicated treatment, separate from its single-agent counterpart.

2.2.1 Flocking and Swarming

The connection between animal behaviors and robotics is rather immediate and pronounced when large groups or teams of individuals are being considered. The elegant and mesmerizing coordinated motions of swarming insects, flocking birds, or schooling fish have repeatedly served as sources of inspiration for multi-robot behaviors, e.g., [40, 70, 115, 142, 190]. In fact, many coordinated motion models, such as Craig Reynolds' celebrated Boids[5] model [356], started out as a scheme for generating artificial, collective animal behaviors, e.g., for the purpose of computer animation.[6] Due to its prominence in the literature as one of the originators of algorithmic multi-agent behaviors, we here, briefly, discuss the Boids model. At its core, Reynolds' model is based on three different multi-agent (or better yet, multi-robot) behaviors, as illustrated in Figure 2.5, namely *separation*, *cohesion*, and *alignment*.

- **Separation:** Move away from nearby agents in order to avoid collisions and crowding agents too closely together.
- **Cohesion:** Move towards the center of mass of neighboring agents in order to keep the team together.
- **Alignment:** Have the agents steer in the direction of the average heading of nearby agents in order to keep the team moving in roughly the same direction.

---

[5]Supposedly, "boid" is how one would pronounce "bird," using a particular New Yorker sub-dialect.

[6]The Boids model made its feature film debut in 1992 in Tim Burton's *Batman Returns*, where penguin armies marched through Gotham City according to the rules of the Boids model [32].

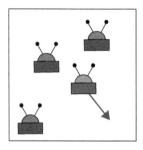  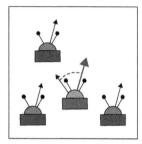

Figure 2.5: Reynolds' Boids: Left: *Separation.* The robot moves away from its neighbors. Middle: *Cohesion.* The robot moves towards its neighbors. Right: *Alignment.* The robot orients its heading towards the average heading of its neighbors.

Of these three rules, the first one (separation) is directly concerned with safety and survival, while the remaining two (cohesion and alignment) are more focused on making the team do something useful beyond just not slamming into each other.

The Boids model's three constituent behavioral building blocks have been augmented, modified, and generalized for a number of different application purposes. In multi-agent robotics, these generalizations (and their biological sources of inspiration) include applications such as foraging (zebras [243]), coverage control (termites [37]), collaborative transport and manipulation of objects (ants [46]), formation control and assembly (birds [324] and fish [97, 399]), and perimeter defense (wolves [256], dolphins [177], and lions [99, 175]—see Figure 2.6).

These types of biologically inspired, coordinated motion patterns are undoubtedly aesthetically pleasing, but they serve other purposes as well in terms application needs. They are also potentially advantageous from a survival vantage point, and ethological studies of animal societies have shown that a collaborative approach offers clear advantages when it comes to achieving a wide range of communal tasks, e.g., [57, 70, 73, 406]. In [406], an extensive catalog of useful, social activities among animals is given, including the following subset:

- Sympathetic induction (i.e., doing the same thing as other individuals, such as yawning);
- Reciprocal behaviors (e.g., feeding young animals through induced regurgitation);

Figure 2.6: Three robotic lions are chasing down a gazelle (Center) using a coordinated pursuit-evasion strategy [99].

- Family and group life behaviors;
- Mating behaviors;
- Persuasion and appeasement;
- Peck orders;
- Antagonistic behaviors (conflict) and fighting behaviors;
- Mutual hostility (spreading the society over a region); and
- Flocking and herding.

Some of these types of social activities have found their way into the multi-agent robotics domain, e.g., [70], where robots are to coordinate their activities based on locally available information, e.g., based on sensor readings. In the following section, we describe how coordinated control strategies can be constructed more generally, and then tie them back to some of the biologically grounded examples and scenarios.

2.2.2 Coordinated Control

Broadly speaking, a distributed multi-robot behavior has to satisfy four different requirements for it to be feasible [93, 253], namely it must be (*i*) *local* in the sense that individual robots can only act on the available information, e.g., through sensing or active communications—this is sometimes referred to as the behavior being "distributed"; (*ii*) *scalable* in that the algorithms executed by the individual robots cannot depend on the size of the entire team—sometimes referred to as "decentralized"; (*iii*) *safe*—as robots are physical agents, deployed in the real world, they must be safe both relative to collisions with each other and relative to the environment; and (*iv*) *emergent* in the sense that global properties should emerge from the local interaction rules—preferably in a provable manner, e.g., [68, 253, 276].

A number of multi-robot behaviors—biologically inspired or otherwise—that satisfy these four requirements have been developed for such purposes as achieving and maintaining formations [197, 311, 318, 438], covering areas [94, 95, 271], securing and tracking boundary curves [277, 397, 440], or mimicking biological, social behaviors, such as flocking and swarming [97, 98, 170, 356]. We here describe how such behaviors can be constructed in a systematic manner.

To make matters concrete and to illustrate how multi-robot behaviors can be produced that satisfy the requirements, consider a collection of N robots, with positions $x_i \in \mathbb{R}^p$, $i = 1, \ldots, N$, with $p = 2$ in the case of planar robots, and $p = 3$ for aerial robots. These robots could, for instance, be equipped with omni-directional range sensors, which enable them to measure the positions of nearby robots relative to their own positions, i.e., Robot i can measure the value $x_j - x_i$ if Robot j is within range of Robot i's sensors.

The flow of information, through sensor measurements or over communication channels, can be encoded abstractly as a neighborhood relation among robots. Formally speaking, we say that Robot j belongs to Robot i's "neighborhood," $j \in N_i$, if information—in some form or another—is flowing from Robot j to Robot i. And, to ensure that the behaviors are indeed local, the robots must act solely based on the available information in such a way that appropriate global objectives are achieved.

Returning to Reynolds' cohesion concept as a starting point, we can define a measure of how far the robots are from each other, i.e.,

$$\mathscr{E}_{cohesion}(x) = \frac{1}{2} \sum_{i=1}^{N} \sum_{j \in N_i} \frac{1}{2} \|x_i - x_j\|^2, \tag{2.4}$$

where we use $x \in \mathbb{R}^{pN}$ to denote the joint state $x = [x_1^T, \ldots, x_N^T]^T$. This performance measure, $\mathscr{E}_{cohesion}$, is zero when the robots are all at the same location, i.e., they have achieved total cohesion, or "rendezvous" [17, 96, 239]. Rendezvous constitutes an extreme form of cohesion—all the robots end up in exactly the same location—but it serves as a fruitful starting point for the discussion about how to generate more useful multi-robot behaviors.

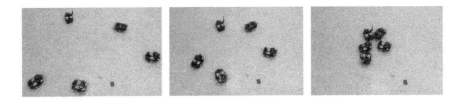

Figure 2.7: A team of robots are solving the rendezvous problem by executing the consensus protocol in Equation 2.6. As prescribed, they end up converging to the same location.

One way of achieving rendezvous, i.e., to make $\mathscr{E}_{cohesion} = 0$, could be to make use of a gradient descent flow, i.e., to let the robots move in the direction of the negative gradient of the cohesion measure. To this end, assuming that the neighborhood relationship is symmetric, the gradient of $\mathscr{E}_{cohesion}$, with respect to the individual robot positions, is given by

$$\frac{\partial \mathscr{E}_{cohesion}(x)}{\partial x_i} = \sum_{j \in N_i} (x_i - x_j), \quad i = 1, \ldots, N, \tag{2.5}$$

giving the negative gradient flow,

$$\dot{x}_i = -\sum_{j \in N_i} (x_i - x_j), \quad i = 1, \ldots, N. \tag{2.6}$$

This motion model is oftentimes referred to as the *consensus equation*[7] [47, 140, 316, 355], and its effect on a robot team is shown in Figure 2.7.

Numerous variations on the consensus equation have been proposed and applied across a wide spectrum of scenarios. For example, let $p = 2$, i.e., the robots are planar, and assume that they are arranged in a "directed cycle topology" such that each robot's neighborhood set is given by the single individual directly ahead of the robot in the cycle. Then, instead of letting the robots "aim" towards their neighbors (following the cohesion recipe), they can instead move with a slight offset, as

$$\dot{x}_i = R(-\psi)(x_{i+1} - x_i), \quad i = 1, \ldots, N-1,$$
$$\dot{x}_N = R(-\psi)(x_1 - x_N), \tag{2.7}$$

---

[7]The reason for the "consensus" monicker is that, by moving around, the robots are agreeing (or reaching consensus) on where to meet.

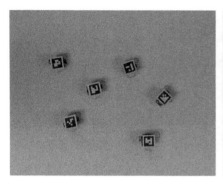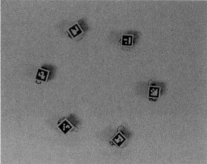

Figure 2.8: Six robots are executing the cyclic pursuit behavior in Equation 2.7, with $\psi = \pi/6$, which leads them to asymptotically converge to a circle formation.

where $R(-\psi)$ is the rotation matrix of angle $-\psi$. If the offset angle is $\psi = \pi/N$, a perfect circular motion is achieved—so-called *cyclic pursuit*—while, if $\psi < \pi/N$, the robots will spiral inwards towards a common location (achieving rendezvous in a somewhat roundabout way), and if $\psi > \pi/N$, they will spiral outwards, away from each other [261]. This is showcased in Figure 2.8.

Similarly, if instead of reaching agreements over the positions, the robots agree on what direction they should move in, i.e., the consensus equation operates on the robot headings instead of their positions,[8]

$$\dot{\phi}_i = -\sum_{j \in N_i} (\phi_i - \phi_j), \quad i = 1, \ldots, N, \tag{2.8}$$

where ϕ_i is the heading of Robot i, then a so-called *flocking* behavior emerges [194, 318, 401], as illustrated in Figure 2.9 (Left). Or, using the terminology established by Reynolds, the robots achieve "alignment" by coming to an agreement over their headings.

From the vantage point of survivability, the consensus equation is a miserable algorithm. It is literally causing the robots to crash into each other, and the end result is an N-robot pileup. To remedy this and turn the consensus

---

[8]This equation must be interpreted modulo 2π as the angles otherwise may do what angles do and wind up with large numbers way beyond 2π, with wildly rotating robots as the result.

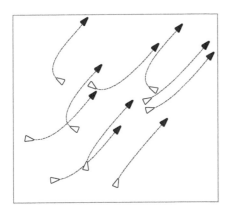 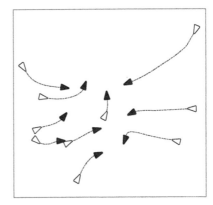

Figure 2.9: Variations on the consensus equation. Left: flocking. Right: formation control. In these figures, the robots start at the lighter locations and arrive at the darker locations. In the formation control scenario, no particular geometric shape is prescribed. Rather, the robots move so as to achieve desired inter-robot distances.

equation into a more useful multi-robot behavior, it needs to be augmented to ensure that the robots do not get too close to each other.

2.2.3 Formation Control

We have now seen examples of cohesion and alignment behaviors, using the terminology borrowed from Craig Reynolds. But what about separation? This is arguably the most important multi-robot behavior for the purpose of safety and, by close association, survival. In fact, as a direct generalization to the cohesion performance measure in Equation 2.4, consider instead the symmetric, pairwise performance cost between Robots i and j, $\mathscr{E}_{ij}(\|x_i - x_j\|) = \mathscr{E}_{ji}(\|x_j - x_i\|)$, where we again assume that the information exchange network is symmetric, i.e., $j \in N_i \Leftrightarrow i \in N_j$.

Gathering up all the terms, the global performance cost is given by

$$\mathscr{E}(x) = \frac{1}{2} \sum_{i=1}^{N} \sum_{j \in N_i} \mathscr{E}_{ij}(\|x_i - x_j\|). \tag{2.9}$$

The Chain Rule tells us that

$$\frac{\partial \mathscr{E}_{ij}(\|x_i - x_j\|)}{\partial x_i} = \frac{\partial \mathscr{E}_{ij}(\|x_i - x_j\|)}{\partial \|x_i - x_j\|} \frac{(x_i - x_j)}{\|x_i - x_j\|} = w_{ij}(\|x_i - x_j\|)(x_i - x_j),$$

$$\tag{2.10}$$

i.e., the partial derivative is a scalar function, w_{ij}, of the inter-robot distance times the relative displacement, $x_i - x_j$. As such, the gradient descent rule not only satisfies the local property required of multi-robot behaviors, it is given by a "weighted consensus protocol,"

$$\dot{x}_i = -\frac{\partial \mathscr{E}}{\partial x_i} = -\sum_{j \in N_i} w_{ij}(\|x_i - x_j\|)(x_i - x_j). \qquad (2.11)$$

The reason why this construction is systematic and theoretically justified is, if we restrict $\mathscr{E}$ to positive semi-definite functions that are 0 only at the desired, global configurations, we note that

$$\frac{d\mathscr{E}}{dt} = \frac{\partial \mathscr{E}^T}{\partial x}\dot{x} = \sum_{i=1}^{N}\frac{\partial \mathscr{E}^T}{\partial x_i}\dot{x}_i = -\left\|\frac{\partial \mathscr{E}}{\partial x}\right\|^2. \qquad (2.12)$$

In other words, $\mathscr{E}$ is a Lyapunov function and one can resort to LaSalle's Invariance Principle, e.g., [209], to ensure that the desired configuration is an (at least locally) asymptotically stable equilibrium point.[9]

A number of examples of this construction have been discussed in the literature. First, as already encountered, the standard consensus equation can be derived from

$$\mathscr{E}_{ij}(\|x_i - x_j\|) = \frac{1}{2}\|x_i - x_j\|^2 \Rightarrow w_{ij} = 1. \qquad (2.13)$$

If the error is just the norm, as opposed to the square of the norm, then

$$\mathscr{E}_{ij}(\|x_i - x_j\|) = \|x_i - x_j\| \Rightarrow w_{ij} = \frac{1}{\|x_i - x_j\|}, \qquad (2.14)$$

which is a form that has been used in [97] to describe coordinated behaviors among schooling fish. The interpretation here is that, as fish pay more attention to nearby fish, the square norm in Equation 2.13 counteracts this by penalizing deviations from faraway fish in an overly aggressive manner.

---

[9]This is true only as long as the neighborhood sets do not change. If they do, e.g., if the robots move in and out of each others' neighborhoods, then $\mathscr{E}$ will experience discontinuities, and either a hybrid version of LaSalle's Invariance Principle must be used, or arguments must be employed that establish that sooner or later, the neighborhood sets become static—see, e.g., [111, 183, 285].

If the robots are supposed to arrange themselves at a prescribed inter-robot distance, δ, we obtain a *formation control* behavior, e.g., [103, 134, 227, 232, 311, 440], as opposed to a rendezvous behavior. An example of this, found in [276], is given by

$$\mathscr{E}_{ij}(\|x_i - x_j\|) = \frac{1}{2}(\|x_i - x_j\| - \delta)^2 \Rightarrow w_{ij} = \frac{\|x_i - x_j\| - \delta}{\|x_i - x_j\|}. \qquad (2.15)$$

The interpretation here is that the weight is negative if the robots are closer than δ apart, thereby repelling each other, while agents that are farther than δ apart are attracted through the corresponding positive weight, as illustrated in Figure 2.9 (Right).

One complication associated with multi-robot networks is that, throughout the maneuvers, the network should typically stay connected for the team to behave properly [197, 371, 437, 439]. Connectivity can be ensured through a *connectivity maintenance* behavior that is constructed to make the weights sufficiently large as the inter-robot distances approach a critical distance, Δ, which is the distance where the robots are no longer able to sense and/or communicate with each other. In [197], the following choice was shown to guarantee connectivity maintenance

$$\mathscr{E}_{ij}(\|x_i - x_j\|) = \frac{\|x_i - x_j\|^2}{\Delta - \|x_i - x_j\|} \Rightarrow w_{ij} = \frac{2\Delta - \|x_i - x_j\|}{(\Delta - \|x_i - x_j\|)^2}. \qquad (2.16)$$

A combined formation control and connectivity maintenance protocol could thus be

$$\mathscr{E}_{ij}(\|x_i - x_j\|) = \frac{1}{2(\Delta - \delta)}\left(\frac{\|x_i - x_j\| - \delta}{\Delta - \|x_i - x_j\|}\right)^2 \Rightarrow$$

$$w_{ij} = \frac{1 - \frac{\delta}{\|x_i - x_j\|}}{(\Delta - \|x_i - x_j\|)^3}, \qquad (2.17)$$

as shown in Figure 2.10. In that figure, the behavior in Equation 2.17 is used to generate a motion pattern that is both provably safe (no collisions) and maintains connectivity (no robots get disconnected from the group). *These two ingredients—safety and connectivity—are going to play significant parts in the broader survivability narrative.*

What all of these constructions show is that it is possible to achieve rich and diverse multi-robot behaviors through a systematic selection of scalar

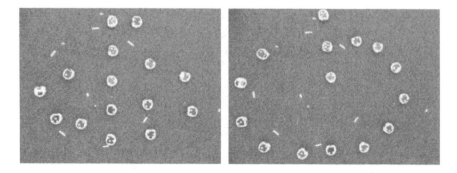

Figure 2.10: 15 robots are forming a "G" by executing the formation control strategy in Equation 2.17.

weights in the consensus equation. But, if the objective is not to assemble a particular shape, but rather to spread the robots out to cover an area, modifications to this construction are needed. This is particularly pertinent to environmental monitoring applications, where robots are to be deployed over a natural domain for the purpose of collecting environmentally relevant data.

2.2.4 Coverage Control

Although coverage control cannot be directly phrased in terms of a negative gradient flow to a sum of pairwise performance costs as in Equation 2.9, it can still be approached through a slightly more involved performance measure. More specifically, the planar coverage control problem concerns itself with having a collection robots cover an area, $\mathscr{D} \subset \mathbb{R}^2$, in an optimal manner [68, 95].

To this end, let each robot be in charge of all the points in $\mathscr{D}$ that are closest to that particular robot. This is not an alien concept to ecology, and in that field, this subset of the domain is known as the agent's (or animal's or animal group's) *region of dominance* [308, 357]. By dividing $\mathscr{D}$ into a collection of such regions, we have produced a partition of $\mathscr{D}$ into N Voronoi cells,[10] with the robot locations, x_i, $i = 1, \ldots, N$, as "seeds,"

$$\mathscr{V}_i(x) = \{p \in \mathscr{D} \mid \|x_i - p\| \leq \|x_j - p\|, \ \forall j \neq i\}. \tag{2.18}$$

---

[10]Note that we use the Euclidean distance to define what it means to be "closest." However, the concept of a Voronoi partition is flexible enough to allow for other notions of distance that can be used to capture robot capabilities, such as limited energy [220], different sensing ranges or footprints [226], or motion constraints [34, 221].

If we assume that points closer to Robot i are covered more effectively by that robot than points farther away, we can associate a strictly increasing cost, as a function of distance, with the quality (or, rather, lack thereof) of the measurements experienced by Robot i, located at x_i, when measuring point $p \in \mathcal{D}$. For the purpose of mathematical clarity, we can let this quality measure be given by the distance squared, $\|x_i - p\|^2$, which, in turn, allows us to write down the so-called *locational cost* associated with the robot positions,

$$\mathcal{E}_{coverage}(x) = \sum_{i=1}^{N} \int_{\mathcal{V}_i(x)} \|x_i - p\|^2 \varphi(p) dp. \tag{2.19}$$

Here, the function $\varphi : \mathcal{D} \to \mathbb{R}$ measures the relative importance of points in the environment, i.e., if $\varphi(p) > \varphi(q)$, then point p is more important than point q for the robot ensemble.

Taking the partial derivative of this locational cost gives

$$\frac{\partial \mathcal{E}_{coverage}}{\partial x_i} = 2 \int_{\mathcal{V}_i(x)} (x_i - p)\varphi(p) dp. \tag{2.20}$$

Following the program laid out in the previous paragraphs, using a gradient descent flow as a way of unleashing LaSalle's Invariance Principle, gives

$$\dot{x}_i = 2 \int_{\mathcal{V}_i(x)} (p - x_i)\varphi(p) dp = 2m_i(x)(x_i - \rho_i(x)), \tag{2.21}$$

where $m_i(x) = \int_{\mathcal{V}_i(x)} \varphi(p) dp$ and $\rho_i(x) = \int_{\mathcal{V}_i(x)} p\varphi(p) dp / m_i(x)$ are, respectively, the total mass (as defined by φ) and center of mass of the ith Voronoi cell.

The expression in Equation 2.21 is a bit untidy. To remedy this, one can scale the control action by a positive gain, to obtain a scaled descent flow. Under the particular choice of gain given by $1/2m_i(x)$, the scaled gradient is given by

$$\frac{1}{2m_i(x)} \frac{\partial \mathcal{E}_{coverage}}{\partial x_i} = x_i - \rho_i(x). \tag{2.22}$$

The resulting, descent-based control law is a continuous-time version of "Lloyd's algorithm" [245], for *coverage control*,

$$\dot{x}_i = \rho_i(x) - x_i. \tag{2.23}$$

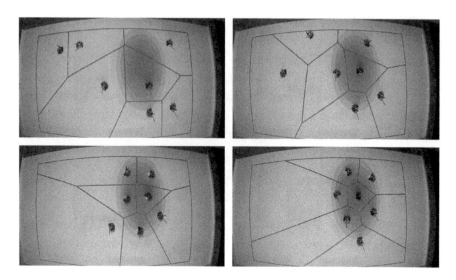

Figure 2.11: By moving against the gradient to the locational cost in Equation 2.19, the robots solve a coverage problem by converging to a Centroidal Voronoi Tesselation. (Points in the environment are shaded according to their importance, as captured by the density function, φ.)

This update law solves the coverage control problem in that it makes the robots reach (asymptotically) a so-called Centroidal Voronoi Tessellation (CVT), whereby $x_i = \rho_i(x)$, $i = 1, \ldots, N$. That this coverage behavior is indeed resulting in an optimal coverage configuration is ensured by the fact that CVTs are locally optimal configurations to the locational cost [95]. An example of this algorithm in action is given in Figure 2.11.

What we have now seen is a collection of valid multi-robot behaviors— from rendezvous and formation control to coverage control—that respect the distributed nature of the robot team, and that can be used to ensure safety and survivability. We will return to these behaviors throughout the narrative as generators of useful multi-robot motions. Just as ecosystems are populated by more than one lonely, single organism, most long-duration deployments will inevitably involve multiple robots. For this reason, it is important to understand not just how to make individual robots safe, but how to ensure safety and survival for the entire team, which is ultimately why it is necessary to establish the basic language for describing multi-robot behaviors. However, the matter of full-blown survival of robot teams will have to wait until subsequent chapters, under the formal umbrella of constraint-based control design.

2.3   The Combinatorics of the Real World

Rather than pursuing additional behavioral design questions, we note that no discussion about robot survival is complete without a nod towards the plethora of objects and events the robots might encounter (and hopefully survive said encounters) during their deployments. The bad news is that long-duration autonomy, taken to its extreme, implies that anything that can happen, sooner or later will indeed happen. For instance, a rule of thumb in the autonomous vehicle literature is that a fatal accident happens in the United States roughly every hundred million miles when cars are piloted by human drivers [195]. As such, for autonomous vehicles to live up to their promise of improving traffic safety, they must exceed that number. But there is a lot of weird stuff that happens on the roads during a hundred-million-mile journey. Cows and deer enter the highways, hailstorms and mudslides erupt, and numerous and miscellaneous toys are tossed out of car windows. In short, it is a strange and messy reality that the robots are expected to cope with. This is what is meant by the *combinatorics of the real world*.

Another way of phrasing this observation is that the number of possible events that might occur, and objects that a robot might encounter during a long-duration deployment, is staggeringly large. And, as a result, any attempt at enumerating all of these possible occurrences and conditions is doomed to fail. This has implications for the robot design in that it must be sufficiently adaptive and resilient to manage despite the lack of a complete catalog of what the world may hold. Because, in long-duration autonomy, *if it might happen, it will happen—sooner or later.*

2.3.1   Elephants Don't Play Chess

Once one embraces the inevitable impossibility of enumerating and modeling everything a robot might encounter during a long-duration journey, one can ask what should be modeled in the first place? Rodney Brooks provided an answer to this question in the paper "Intelligence Without Representation" [64], where he stated, "The world is its own best model." The point is that robots need not (or even stronger—should not) use elaborate, symbolic models to represent the environments in which they are embedded. In fact, the title of this section, "Elephants Don't Play Chess," is borrowed verbatim from the title of another one of Brooks' papers [63] that investigates what implications this simple yet impactful observation about "situatedness"[11] might have for the design of robot architectures.

Without staying particularly faithful to the discussion in [63], but rather mapping its ideas onto the realm of long-duration autonomy and robot survival, one can start by considering what the actual game of chess is all about in order to understand why it is perhaps not overly popular among elephants. The chess board itself is comprised of 64 well-defined squares of alternating colors, corresponding to the distinct locations where the pieces can be. Each type of piece has a finite set of moves that it is allowed to execute, as a way of moving from one square to the next. And, the goal of the game is clear and well-defined—to checkmate the opponent's king piece. Of course, this curt description completely misses the mark in terms of describing a beautiful and maddeningly complex, strategic game. But it calls out a number of salient features that may make it a poor starting point for understanding robots deployed in a dynamic environment over long time-scales. After all, elephants (and perhaps, by analogy, robots brought up under the robot ecology paradigm) don't play chess.

First, what is a "square" out in the real world? When elephants (or people or other animals) move around, it is not done by somehow transitioning between discrete cells in a grid. Instead, the world is a continuous and, as noted, oftentimes quite messy place. Without simplifying abstractions and assumptions that allow us to think of the world as discretized and gridded, the motion of the animal (or robot) must be anchored and understood in this frame of reference. Second, the notion of a "move," drawn from a finite set of allowable ways of transitioning through the environment, is rather unclear. In the world of Newton, quantities like velocities and accelerations make sense. But arguably, a motion pattern involving moving two squares in some direction followed by one square in an orthogonal direction—as would a chess knight—does not really fit the bill. And, third, the idea that there is some ultimate "goal" that the animal should try to achieve in order to win the game of life is also problematic. Without getting into a discussion about the Meaning

---

[11] Brooks used the term "situatedness" to explicitly call out the fact that the robot behavior only makes sense when situated in an environment [64]. This is a clear and direct parallel to robot ecology.

of Life, this notion of an overarching goal is certainly not a straightforward concept.[12]

So if elephants don't play chess, what do they do? Well, they survive (or at least try to). This is not overly surprising, since, as already pointed out, survival is a prerequisite to thriving, and the first order of business when designing robots and robot architectures for long-duration autonomy must be to elucidate this point. In fact, [284] goes so far as to state that "effective mobility, acute sensing, and survival constitute the prerequisites for intelligence." These observations were formalized in [64], where Brooks asked what an engineering methodology for building "Creatures" would entail, and the following principles were found:

- "A Creature must cope appropriately and in a timely fashion with changes in its dynamic environment."
- "A Creature should be robust to its environment; minor changes in the properties of the world should not lead to total collapse of the Creature's behavior; rather one should expect only a gradual change in capabilities of the Creature as the environment changes more and more."
- "A Creature should be able to maintain multiple goals and, depending on the circumstances it finds itself in, change which particular goals it is actively pursuing; thus it can both adapt to surroundings and capitalize on fortuitous circumstances."
- "A Creature should do something in the world; it should have some purpose in being."

The resulting architectural design, proposed in [64] for realizing this methodology, was the *Subsumption Architecture*, whereby different "layers" (akin to behaviors for the purpose of this discussion) are working on different goals in a concurrent and asynchronous fashion. External stimuli are then used to inhibit or suppress the outputs from the different layers, based on a priority-based arbitration mechanism, where higher priority behaviors/layers would take priority over (or subsume) lower priority ones.

Within the survival themes of this chapter and the broader robot ecology motif, these architectural ideas are quite promising. The recurrent observation that timely responses to dynamic effects is necessary certainly seems like

---

[12] A biologist would probably say that such a goal might be phrased in evolutionary terms, e.g., involve statements about maximizing fitness for the purpose of natural selection.

a must-have when ensuring robot survival. Similarly, it seems rather reasonable that the closer the robot gets to a potentially catastrophic situation, the more consideration should be given to trying to avoid that from happening. Finally, the need to manage a rich and complex environment without being able to enumerate and symbolically represent everything the robot might encounter in its environment also points towards something that is highly desirable in a long-duration architecture. We summarize these three key points below:

> Survival is a prerequisite to thriving in dynamic environments, which implies that the control mechanisms must satisfy these condition:
>
> 1. The robot's response to its sensor measurements must be sufficiently fast relative to the time-scale of the environment;
> 2. The closer one gets to a catastrophic failure, the more urgency and care must be given to the avoidance of this situation; and
> 3. Safety and survival cannot rely on explicit enumerations and symbolic models of everything the robot may encounter in the world.

2.3.2 Technology Readiness Levels

An alternative way of managing the combinatorics of the real world is to try to avoid facing them altogether by carefully organizing and controlling the environmental conditions the robot might encounter. This approach may work when the deployment happens over short time horizons (relative to the environmental time-scales), where the conditions remain largely constant, in order to cut down on their potential variability. For example, a computer vision algorithm carefully calibrated and optimized for a particular outdoor lighting condition is probably fine over a few minutes, but certainly not if the deployment spans multiple hours or days. Alternatively, the variability can be kept low artificially, e.g., by maintaining constant lighting conditions in an indoor environment, such as warehouses or manufacturing facilities, no matter what time of day, or even year, it is. Neither of these two approaches are particularly insightful for the purpose of robot survival and long-duration autonomy since they are simply dodging and, as such, ignoring one of the salient complications with unknown and significant environmental variability.

There is, however, a possible middle ground between the daunting requirement to cope with highly varying and unknown conditions and carefully curating and controlling the environmental deployment conditions. This middle ground would involve designing the system incrementally in such a way that only once it is functioning robustly for a given set of restricted and limited-variability environmental conditions, these conditions are allowed to become slightly more complex and vary in a controlled fashion. The system itself will then have to be augmented and made more robust in lockstep with each such increase in environmental complexity and variability.

A prime example of this methodology is NASA's Technology Readiness Levels (TRLs), consisting of nine levels [181], where TRL 1 is basically a pen-and-paper conceptual design, while TRL 9 is robust deployment on Mars (or wherever the technology deployment is targeting). More specifically, the NASA scale [293] is

TRL 1: Basic principles observed and reported

TRL 2: Technology concept and/or application formulated

TRL 3: Analytical and experimental critical function and/or characteristic proof-of concept

TRL 4: Component and/or breadboard validation in laboratory environment

TRL 5: Component and/or breadboard validation in relevant environment

TRL 6: System/subsystem model or prototype demonstration in a relevant environment

TRL 7: System prototype demonstration in a space environment

TRL 8: Actual system completed and "flight qualified" through test and demonstration

TRL 9: Actual system "flight proven" through successful mission operations

Of course, NASA, as well as organizations that subsequently adopted the TRL formalism for estimating a technology's maturity, did not formulate the TRLs in terms of survival or incremental increase of the environmental variability. Instead, the readiness levels describe the maturity across different dimensions, such as program concepts, specifications, demonstrated capabilities, and technology-operational environment match [110]. Of these, the most relevant maturity concept to the discussion about robot ecology is the explicit focus on the technology's relevance to its intended "operational environment" and the idea that although a technology can be highly mature in one environment, the mismatch might be significant in another. And, to transition up the TRL-ladder, this technology-environment fit must be explicitly demonstrated under increasingly demanding specifications.

In the context of survivability and long-duration autonomy, as promoted here, the NASA TRL scales are not immediately applicable. They do, however, point out how one can design robot technologies and systems in a disciplined manner to achieve long-duration autonomy thanks to ruggedized, redundant, and over-all clever engineering solutions. After all, the book started by following in NASA's footsteps with the two remarkable Mars rovers, *Spirit* and *Opportunity*. The approach taken here, however, is to try to elucidate fundamental design principles for why long-duration autonomy must be approached differently as compared to its short-duration counterpart.[13] For example, the idea of increased environmental variability, as expressed in the TRL framework, is highly relevant to robot safety, as we will see in a subsequent section, where robots learn how to safely engage with their surroundings by incrementally increasing the size of the safe set as they gain new competencies. This, in turn, allows the robots to explore previously unavailable state-action pairs for the purpose of furthering their capabilities.

2.3.3 Constraints and Laws of Robotics

We have now seen a number of approaches to robot architectural design, where immediate concerns are prioritized more highly than less urgent tasks as a way of reacting prudently and timely to dynamic, environmental factors. But, if we, somewhat pompously, elevate *survival* to the First Law of Long-Duration Autonomy, something stronger is probably needed than prioritization strategies based on weighted combinations of robot behaviors, as was the case with a number of arbitration mechanisms.

To draw additional inspiration and frame the discussion in a more popular context, consider another First Law, namely Isaac Asimov's First Law of Robotics [27],

A1: A robot may not injure a human being or, through inaction, allow a human being to come to harm.

---

[13]We note that it is not entirely without irony that this book is principally a TRL 1 inquiry into what a TRL 9 solution looks like for the problem of long-duration autonomy.

We immediately note that Asimov's First Law does not say that not harming humans should be given higher priority than other tasks. It says that robots should not harm humans at all. Period! One could of course interpret this as giving complete and total control to the highest priority behavior (do not harm humans), i.e., to switch control authority among behaviors and only consider the particular behavior that is currently most urgent or important. But why not chew gum and walk at the same time? Why not let the robot do multiple things concurrently? That would certainly make it a more useful robot. Also here Asimov can provide some literary guidance through his Second Law of Robotics:

A2: A robot must obey the orders given it by human beings except where such orders would conflict with the First Law.

The combination of these two "laws" has the semantics of "Do X while ensuring that Y does not happen" or, phrased slightly differently, "Solve a given task (obey human orders) as well as possible, subject to the constraint that something else (harming humans) is ensured never to happen." While fully acknowledging that this is derived from the world of science fiction, it is still instructive to replace "not harming humans" with "survival." This modification takes us to a construction, hinted at already in the introductory chapter, where survival takes on the form of a *constraint*.

Staying within the realm of Laws of Robotics, there are a handful of other laws that point in the same direction, i.e., that the primary consideration for a robot is phrased in terms of a constraint.[14] For example, Mark Tilden proposed a different set of laws [174], and the first two of Tilden's Laws of Robotics state that

T1: A robot must protect its existence at all costs.
T2: A robot must obtain and maintain access to its own power source.

These two laws are even more well-aligned with the survival theme, as the first of Tilden's Laws is literally saying that whatever the robot does, it must do it subject to a survival constraint. And, as an additional forward-looking nod to the *persistification* idea in Chapter 5, the robot should never completely deplete its batteries. And, making sure that the "tank" never becomes empty is part of the robot's own set of responsibilities that must be managed autonomously as an integral part of the control design.

---

[14]For a less literary and more scientific approach, the rich literature on robot ethics has a related and sometimes similar take on these issues, e.g., [240, 241].

The main punchline with these informal yet illuminating observations is that the semantics of constrained optimization lends itself naturally to describe Laws of Robotics as well as what long-duration autonomy is all about, namely:

> *minimize:* some performance objective
>
> *subject to:* don't die

As we will see in the next chapter, this is also an ecologically consistent way of formulating the design problem. But, before we can explore these ecological connections, we first outline a few, promising directions that are immediately opened up by this constraint-based design framework.

Safe Learning

The survival constraint—don't die—is situated, to borrow the term from [64], in the sense that it only makes sense relative to how the robot's capabilities match a given environment. A puddle is not a deal-breaker to a waterproof robot, and a pothole is not a big deal to a robot that can jump. For other robots, these minor inconveniences and obstructions can constitute fatal, existential threats. What this means is that the connection between environment and robot (or, robot dynamics, to be more specific) is key to survivability.

But what if the robot does not yet know how to jump (or even if it can jump at all) over holes, but wants to learn? The traditional approach to machine learning in general, and reinforcement learning in particular, would be to have the robot try a bunch of hop-like motions in the vicinity of all kinds of holes, i.e., it would explore all state-action pairs for the purpose of learning, and would associate rewards with certain actions in given situations, e.g., [10, 202, 288, 398]. But that would be a recipe for disaster in a long-duration context. Sooner or later the robot would perform a move that would result in it falling to its doom.

What is needed is a formal notion of *safe learning* that restricts the learning process to only safe configurations. Initially, the safe set, as construed through the survival constraint, needs to be small, as the robot has not yet learned much about its capabilities and how well they match its environment. As it gains in competency by learning more about how it can interact (safely) with the environment, this set can be expanded, encompassing more state-action pairs, e.g., [5, 28, 45, 426]. This is happening all the while ensuring

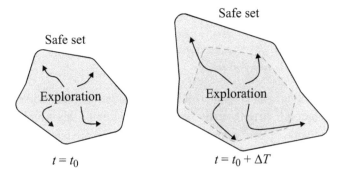

Figure 2.12: Initially, at time t_0, the safe set is small as the robot does not know much about how to move safely through the environment. Over time, this set expands, allowing for a more adventurous approach to the learning enterprise.

that the learning process remains safe, given the current state of the survival constraint, as depicted in Figure 2.12.

This program, with clear connections to NASA's TRL paradigm, was investigated in [426], where a small quadcopter learned how to fly through a blade-less fan without getting blown out of the air. By probing the airflow around the fan in an increasingly daring (yet safe) manner, the quadcopter learned to fly through the fan, as discussed in technical detail in Chapter 4 and displayed in Figure 4.12. This is a feat it would not have been able to accomplish without restricting the learning process to only explore an expanding, safe set.

Curiosity-Driven Learning

If the First Law of Long-Duration Autonomy is the satisfaction of a constraint that says something along the lines of "Don't die!" then the Second Law could perhaps be "Do something useful!" as long as the First Law is not violated. But what should this useful pursuit be? The answer clearly depends on what the robot has been tasked to do. However, one can certainly envision situations, in between tasks, when the robot is not tasked to do anything. Those situations seem like golden opportunities for the robot to learn new things. If the robot is not tasked with anything in particular—as it is in between jobs— then the reward-function that typically drives the learning process must come from somewhere else.

Curiosity seems like a good place to start when looking for a driver of the learning process. Merriam-Webster [274] defines curiosity as

Curiosity

1: desire to know
> *a: inquisitive interest in others' concerns*
> *b: interest leading to inquiry*

Of those two definitions, 1b—interest leading to inquiry—seems like the pertinent interpretation.[15] In the machine learning community, this is typically taken to mean that the tasks (and, consequently, the rewards) are intrinsic rather than extrinsic. In other words, the robot itself has to generate the tasks ("interest") and then set out to learn how to perform them ("leading to inquiry").

One route, as pursued in [217, 416], is to let the tasks be more or less randomly generated. But that approach does not gel particularly well with the safe learning paradigm, as previously discussed. Another route is thus to purposefully explore mismatches between modeled and actual capabilities, as was suggested in [393]. This perspective is more in line with the survival theme. What one can envision is a learning process that is explicitly targeted at exploring parts of the safe set that are poorly explored and understood. As a consequence, one arrives at an approach where curiosity-driven learning and safe learning come together under a unified framework for what robots with free time should spend their time doing, beyond merely surviving.

Autonomy-on-Demand

Given the proposition that if there was such a thing as a Second Law of Long-Duration Autonomy, it would say something like "Do something useful!" albeit perhaps stated a bit more eloquently.[16] What does this actually entail? We have already seen at least three different interpretations of this second law, namely (*i*) to minimize some mysterious performance cost, (*ii*) to learn new skills, and (*iii*) to do nothing at all in order to preserve energy. Of these three possible answers, the latter two are reasonably clear. But what about the first option? Where, exactly, is this performance measure originating?

Imagine a team of robots, deployed in some natural environment for long periods of time, where they spend most of their time doing nothing at all. Every now and then, they interrupt their leisurely existence and start exploring their surroundings and capabilities, a transition that is triggered by "curiosity"

---

[15]The other definition is quite intriguing as well, and the idea of nosy robots that won't mind other robots' affairs is, at least, amusing. It could even be interpreted as having implications for *imitation learning*, where robots learn by explicitly having an active and "inquisitive interest in others' concerns" [33, 369].

[16]Recall that this was also one of the principles for Brooks' Creature design [64].

rather than by external factors. The exception to these two activities is that, every so often, the robots move to particularly sunny areas to recharge their batteries. What the team is really doing is inhabiting an ecosystem, waiting to be recruited. And in between deployments, they learn new skills. The way the robots get recruited is through the introduction of a new cost function, i.e., they are given new, user-specified instructions about how they should interpret "Do something useful!"

We dub this set-up "autonomy-on-demand." It supports taskable and recruitable robots that are part of a broader ecosystem. The robots can improve their capabilities over time. And, no matter what, they adhere to the (somewhat grandiose) First Law of Long-Duration Autonomy, namely "Don't die!"

3 Ecological Connections

Now that we have seen how "situatedness" [64] matters to robotics from safety and survival vantage points, it is worth taking a detour into the world of ecology, which deals exactly with this topic, i.e., a systematic *study of habitats and the organisms that reside within them* [308, 357, 385]. Etymologically, ecology comes from the Greek word "oikos," which means "house," and just as people (and, increasingly, robots) reside in houses, organisms—which constitute the fundamental, indivisible units of ecology—reside in natural environments.[1] It is almost entirely pointless to remove the organism from its environment when trying to understand the intricacies and interconnections driving its behavior.

To clearly call out the difference between organisms in isolation and organisms coupled to their environments, the distinction is made between the *genotype* of an organism, which is the genetic characteristics of the organism, and the *phenotype* that results from the genotype interacting with the environment. The latter encompasses not only the organism's behavior, but also its physical appearance, e.g., [109]. Connecting back to the previous chapter, the "genotype" of the robot could be its hardware configuration and the

---

[1]The irony is not lost on us that in order to understand robots in natural environments, e.g., *not* in houses, we need to study a field whose name literally is derived from the word "house."

Figure 3.1: The robot from Figure 1.4 is learning, over time, to predict "terrain preferability" by interacting with the environment (literally, by driving into things). Left: Camera image. Right: Classified cost of preferable terrain, where lighter terrains are classified as more preferable from an ease-of-navigation vantage point.

software that is initially loaded onto the robot when it embarks on its long-duration deployment. The "phenotype," meanwhile, is the resulting set of behaviors that arises from how the environmental factors are perceived and acted upon by the robot—both through its sensor suites and perception routines, but also as outcomes from adaptive strategies, such as the previously discussed, curiosity-driven learning processes, whereby the robots learn how to interact with their environments.

We have already seen an example of this phenotype idea, on display in a robotics setting in Figure 1.4, where a robot learns, over time, to associate color and texture features, obtained by a monocular camera, with its ability to traverse the terrain [396]. This mechanism is further illustrated in Figure 3.1. By adaptively learning the "terrain preferability," i.e., how well the robot can navigate and traverse different types of terrains, the behavior of the robot is quite different after it has resided in the environment for a while (phenotype) as compared to how it initially operated (genotype).

Individual organisms of the same species within a habitat make up a population, and the highly robotics-relevant notion of survival is one of the two main factors that determine the size and success of the population. When the mortality rate goes down, the population size tends to grow. This is a concept that makes immediate sense in long-duration settings, where survival is a prerequisite to thriving. The other primary contributor to a population's success, *fecundity*, does not quite make as much direct sense in a robotics context, as fecundity is the ability to produce an abundance of offspring. Robotics aside, what we have arrived at is a central, ecological principle, namely that change in population size equals births minus deaths.

In this chapter, we will discuss some of the key ecological concepts and principles needed for the formulation of the robot ecology framework. It is not to be taken as an all-encompassing treatment of the vast discipline of ecology. Instead, it should be understood as an opportunistic path through this vibrant, intellectual field for the purpose of understanding how to think about robot design, when the robots are to be deployed in environments over truly long time-scales. The aim, however, is ultimately *not* to mimic ecology. Rather, it is to arrive at a set of sound, organizing principles for *robot ecology*.

3.1   Organisms and Environments

As ecology—and robot ecology too, for that matter—is the study of organisms in their environments, and because the connection between environment and organism is critically important to the understanding of behaviors, population sizes, and adaptation, this chapter starts with a brief discussion about how to make sense of this connection, all the while pointing forward towards design principles for long-duration autonomy. Environmental factors that matter to animals and robots alike include things like weather and climate conditions, e.g., temperature, humidity, and salinity, as well as the availability of *resources*, which cover a wide range of factors, such as the prevalence of food sources and suitable hiding places. Some resources are intrinsic to the environment (like hiding places), while others are extrinsic (like the availability of prey).

3.1.1   Consumers and Resources

Broadly speaking, environments and their inhabitants can be divided into two categories, namely *consumers* and *resources* [308]. It is, of course, possible for an individual to be both at the same time; zebras are consumers of grass, but are resources for lions. The success of a consumer in a particular environment is directly influenced by the availability of its preferred resources. Or, phrased a different way, the availability (or lack thereof) of resources imposes constraints on what constitutes viable consumer behaviors as well as on how many consumers can be supported by the environment.

When trying to establish what constraints the environment imposes on a population in terms of the size that can be supported, the logistic equation conveniently abstracts away the particulars of the environmental factors through the scalar constants, r (the rate of growth supported by the environment—known as the "inherent per-capita growth rate"), and K (the "carrying capacity," i.e., the number of individuals in the population that can be supported at equilibrium) [26, 331]. If N is the population size, the logistic equation states that

$$\dot{N} = rN\left(1 - \frac{N}{K}\right),$$ (3.1)

whose solution is given by

$$N(t) = \frac{K}{1 + be^{-rt}},$$ (3.2)

where $b = (K - N(0))/N(0)$, as illustrated later in the chapter in Figure 3.6 (Top Left).

The solution to the logistic equation in Equation 3.1 converges asymptotically to the carrying capacity. Although the logistic equation captures some of the intricate consumer-resource relationships present in an environment, it does assume that all individuals have equal and unrestricted access to the available resources, i.e., no "crowding" effects kick in. But, at high densities, individuals within the same species do get in each others' ways as they are attempting to utilize the same resources.[2]

These types of crowding effects are observed also in robotics, e.g., [268, 269]. In particular, for multi-robot *foraging* [40, 70, 77, 375] and *collection* tasks [39, 264], it has been shown that while adding more robots to the group at intermediate robot densities increases the total number of objects found and collected by the team as a whole, this increase in performance is in fact sublinear, caused by a gradual decrease in the individual robot performances, e.g., [180, 234, 321]. Owing to the (typically linear in the team size) operational costs associated with deploying large robot teams, one pertinent question thus becomes: How many robots should be used in order to obtain the desired, overall productivity, while utilizing as few robots as possible?

One possible clue to the question of how many robots are needed can be found through the study of social insects. Entomological studies have shown that many social insect colonies possess the ability to regulate interference

---

[2]"High" densities must be interpreted in a relative sense—different organisms have different densities at which the crowding effects kick in.

using quite simple and decentralized techniques, e.g., [4, 115, 125, 148]. In [4], for example, it was shown how a certain type of ant (the *Solenopsis invicta*) can regulate the densities in narrow tunnels and, subsequently, prevent the formation of flow-stopping clogs. This regulation mechanism is attributed to the propensity of individual ants to reverse out of over-crowded tunnels, thus reducing the overall density. In other contexts, it has been shown that ants use similar inter-ant encounters as the primary sensory mechanism to regulate their densities and perform functions like task allocation and division of labor [165]. Inspired by these observations about density regulation among ants, [267] showed how crowding effects can be leveraged in robotics for the purpose of effectively performing collection tasks, where the robots leave an area of interest if the locally detected encounter rates become too high, as shown in Figure 3.2.

A multitude of similar models have been proposed and developed for describing population-level dynamics and environmental interactions of varying degrees of complexity, sophistication, and fidelity. Although these models are interesting and relevant, our focus will ultimately have to be on robot design principles, which will inevitably involve designing control programs for *individual* robots. As such, we must shift our attention to individuals rather than treating populations as a whole.

Making this shift from populations to individuals, we must ask, how does the consumer-resource relationship manifest itself to an organism inhabiting a particular environment? When an animal moves through a habitat in search of food, it encounters prey throughout its quest. And every time it does, a decision must be made whether or not to pursue the prey. Each such decision corresponds to an investment in terms of time and energy, and the outcome of the decision is not given. One possible outcome is that the prey is subdued and, subsequently, eaten, which typically makes the energy investment worthwhile. But another possible outcome is that the hunt is unsuccessful, and precious assets are expended with nothing to show for it. On top of that, there is an opportunity cost associated with this choice, regardless of the outcome. Potential, not-yet-encountered prey are passed up as a consequence of the decision to pursue this particular snack.

So what is an animal to do? The choice it is facing seems like a highly complex decision problem, and reasoning under uncertainty is not an easy task. As observed in [79, 254], if the animal encounters prey rarely, and it is relatively easy to subdue and consume the prey, then the best strategy is to try

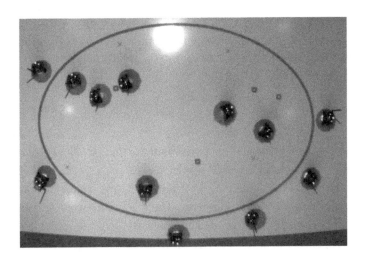

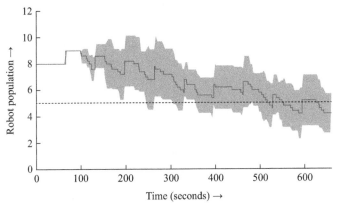

Figure 3.2: Top: Twelve robots execute the "voluntary retreat" algorithm in [267], where the robots decide whether or not to participate in a task by measuring how many local robot-robot encounters they experience. If they experience too many such encounters, they simply leave the area (denoted by the ellipse) and are no longer participating in the task. Robots outside the area get back in by following a probabilistic rule. Bottom: The average robot population size inside the area is plotted across a number of experiments, along with the standard deviation (shaded). As can be seen, the population size approaches the analytically established, optimal value (dotted), using a simple robot-robot interaction rule, akin to the ant-ant interaction rule in [4]. Source: Fig. 6 in Mayya et al. (2019).

to eat every prey (and every viable type of prey) that it comes across. If the opposite is true, i.e., the animal encounters prey quite often and the energy investment required to catch and eat its food is significant, it should be much more of a picky eater, and be highly selective about trying to catch its next meal.

To complicate matters, whatever policy the animal settles on, it cannot be static, as conditions and environments vary over time. To elucidate this point, consider the concept of *environmental patches*. If the organism is the fundamental unit in ecology as a whole, the patch plays this role in landscape ecology [413]. It is defined as a relatively homogeneous area that is somehow distinct from its surroundings. The patch also evolves dynamically as a single unit in that fluctuations and changes occur across the patch as a whole, through what is known as "patch dynamics" [299]. And, animal foraging strategies must keep up with the patch dynamics. This is true also for robots in that strategies that may work well under particular lighting conditions, on flat ground surfaces, or in tight temperature ranges, may fail under different conditions [392].

As a predator is searching a patch for prey, it may initially encounter food at a relatively high rate. But, these resources may deplete over time as they are literally being consumed. This means that at some point, it is worth the investment for the predator to leave the patch and move on to another one. But, this decision depends on a number of parameters like the current patch quality, the expected patch quality of the new patch, the travel time between patches, and the overall risk associated with the move. So even though "the grass is always greener on the other side of the fence," the decision to pursue this verdure depends on how much greener it is compared to the current grass, and on how tall the fence is.

Complicating the deliberation process further is the fact that frequently many different individuals face the same choices and options. As such, the optimal choice is influenced by the decisions made by other members of the population. Under the assumptions that (*i*) animals have perfect knowledge of each patch's quality, and (*ii*) they can move, unimpeded, between patches, the result of the deliberation process is that each individual in the population will select and exploit a patch of equal "realized quality," regardless of the patches' intrinsic qualities [330, 345]. As the first assumption is *ideal*, and the second describes *free* movement, the resulting distribution of individuals across patches is known as the *ideal free distribution*, which is an attribute of animal space utilization first recorded in the context of territorial birds [154].

This idea of equal realized quality resonates quite well with the coverage control construction in the previous chapter. In particular, when defining the locational cost in Equation 2.19, one key component was the density function, $\varphi : \mathscr{D} \to \mathbb{R}_+$, that associates a relative importance to each point in the environment, $\mathscr{D}$, from a monitoring point of view. But, one could just as well let the robots be consumers, and let the density function be a distribution of resources over the environment. Under this interpretation, the amount of resources available to Consumer i (formerly known as Robot i) is given by

$$m_i(x) = \int_{\mathscr{V}_i(x)} \varphi(p)dp, \tag{3.3}$$

where $\mathscr{V}_i(x)$ is the i th Voronoi cell, i.e., the region comprised of all points in $\mathscr{D}$ that are closer to Consumer i than to any other consumer.

In Chapter 2, we saw that by moving against the gradient to the locational cost in Equation 2.19, the robots achieve effective coverage of the environment in a distributed fashion. This may not necessarily be how real animals behave (in fact, they probably do not), but it is a convenient mathematical construction that enables well-performing multi-robot teams.[3] As the individuals execute the decentralized coverage control algorithm in Equation 2.23, they end up at a Centroidal Voronoi Tesselation, which, although not necessarily a configuration where all the cell masses, m_i, are exactly the same (which would translate to the same amount of resources in each cell, i.e., equal realized quality), they are locally optimal in the sense that small variations in robot positions will not improve the overall value of the locational cost. Moreover, as established in the *hexagonal theorem*[4] [121, 122], when the density function is uniform across the entire domain, the cells in the Centroidal Voronoi Tesselation are hexagonal and of equal size, except at the boundary of the domain, as seen in Figure 3.3 (Right).

3.1.2 Niches and Fitness Sets

Besides making decisions about optimal environmental patches and whether or not it is worth the effort in terms of energy and time to pursue a particular prey, what other choices are consumers faced with as they go about their daily

---

[3]The idea of connecting multi-robot behaviors to generalizations of an ideal free distribution is pursued in detail in [330, 345].

[4]This result is only proven in two dimensions, and it is known as "Gersho's Conjecture" in higher dimensions [123].

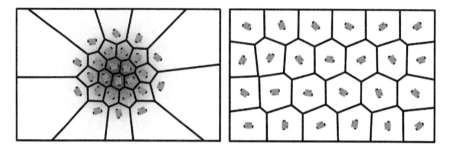

Figure 3.3: 25 robots are executing the coverage control algorithm in Equation 2.23, converging to a Centroidal Voronoi Tessellation, where the mass from Equation 3.3 in each Voronoi cell is equal across all cells, thus achieving something akin to what happens in nature, following the ideal free distribution, i.e., achieving equal realized resource quality. Left: Resource distribution given by a spatially centered Gaussian density function. Right: Uniformly distributed resources across the domain.

(or nightly) business? Ultimately, they must balance the allocation of time and constrained resources to a wide variety of competing functions. For example, time spent searching for food cannot be used to care for offspring or to watch out for predators. There are a number of inherent tradeoffs that must be made. And, at the end of the day, they all come down to considerations about *energy* in one form of another.

We, already in the introduction, featured the poignant quote from [357] that stated that living systems are defined by a "purposeful expenditure of energy." The available energy budget can be spent on such high-level pursuits as migrating, mating, or pursuing prey. In addition to these behavioral and external manifestations of the energy calculus, the maintenance of basic bodily functions requires an energy investment as well. Many different strategies to meet this challenge are on display in nature. The question of how to spend precious energy resources is faced also by robots on long-duration deployments. Movement, for the purpose of satisfying some mission-centric task, must be undertaken only if there are ways of replenishing the batteries. Internal operations, such as running computationally intense and costly deep neural networks, or computing optimized, stabilizing gait trajectories for bipedal, humanoid robots, come at significant energetic costs as well, e.g., [205, 244, 354].

Homeostasis refers to the ability of an individual to maintain constant internal conditions, such as body temperature, water content, glucose and mineral levels, or blood pH, in the face of varying, external environmental conditions [145, 353]. Not all animals rely on homeostasis, as it does indeed require energy expenditures, and those that do are fittingly known as "regulators." For instance, mammals regulate their core temperatures to a narrow band through both physiological thermoregulation (internal, chemical means) and behavioral thermoregulation (also known as "allostasis," where animals respond to hot temperatures by seeking shade or reducing their activity levels and respond to cold temperatures by huddling together or increasing their activity levels).

Behavioral thermoregulation normally takes precedence over physiological, as it is typically faster acting and can accommodate a wider range of temperature swings [145]. Certain animals, such as frogs or turtles, rely much more heavily on these behavioral strategies, as their ability to generate physiological sources of heat is limited. Other strategies to cope with the temperature regulation problem is to dramatically alter the metabolic rate under some circumstances, e.g., through hibernation (a strategy employed by bears and bats, among others), or to be constitutionally able to tolerate wider ranges of internal temperatures. Examples of animals that possess this ability include fish, amphibians, and reptiles, as well as many invertebrate animals. Even some mammals, such as the naked mole-rat and our inspirational, low-energy lifestyle species—the sloths—can tolerate a relatively wide range of internal temperatures [106].

The point with this discussion is that although energy availability is the ultimate, ecological (as well as robot ecological) constraint, the response to that constraint is quite varied, and there is not a one-size-fits-all way to meet this challenge. Although energy may be the most relevant and limiting factor in the robot ecology paradigm, animals have to cope with, and juggle, a number of other critical, limiting factors, such as nutrients, light, temperature, and rainfall. With these considerations in mind, and recognizing that the world is not constant and uniform, it is quite clear that a single genotype cannot, almost by definition, be best suited to all environmental conditions, which leads to the ecological notion of *fitness under variable conditions*, and the idea of an *ecological niche* [81].

To appreciate the notion of "fitness under variable conditions" in action, let us take a quick detour through the lives of our old acquaintances, the sloths. Shade-grown agriculture, employed for example in cacao plantations, has been proposed as an effective strategy for conserving tropical biodiversity. In [332], this practice was investigated in the context of its impact on the local sloth populations, and it was found that shade-grown cacao enterprises were able to sustain healthy populations of two-toed sloths (*Choloepus hoffmanni*), but not so for the brown-throated, three-toed sloth (*Bradypus variegatus*). One can then ask what the differences are between these two species (beyond the number of toes)? The findings in [332] indicate that the two-toed sloth is able to survive, and even thrive to a certain degree, in a number of different environments, while the three-toed sloth is intimately linked to an existence in a particular, preferred tree, namely the cecropia tree, also known as the sloth-tree. In the absence of those trees, the three-toed population was simply not able to sustain itself at any significant levels. Another way of putting this is that the two-toed sloth is more of a *generalist* (relatively speaking), while the three-toed sloth is a *specialist*. As a result, the two-toed sloth exhibits greater fitness under variable conditions, as compared to its three-toed counterpart [328, 332].

In ecology, the idea of variable fitness is captured through the concept of the "fitness set" [380], which denotes the habitat patches where a particular phenotype is best suited over other phenotypes. The greater this set is, the greater fitness under variable conditions that phenotype exhibits. This notion is further illustrated in Figure 3.4, as the fitness set concept is important also in robotics. A robot whose fitness set is given by the particular, highly curated laboratory conditions in which it was debugged and carefully calibrated, is highly unlikely to do well once moved to a different environment. One design principle for robot ecology, that will have to be formalized in subsequent chapters, is thus to strive for control strategies that maximize the robots' fitness sets.

But what makes a particular phenotype well-suited for a given environmental patch? The local abundance of certain species in an area is an indirect yet illuminating reflection of how the individuals belonging to those species take advantage of, and utilize, the available resources. Common species consume a bigger share of the resources as compared to more rare species, which could mean that the individuals belonging to the more common species are able to utilize a broader variety of resources, i.e., they are generalists, like the two-toed sloth. Or, they can still be specialists, but the particular resources they specialize in are abundant; if all you have is a hammer, and the world around you is filled with nails, then you are well-positioned for success. The third possible reason for the relative abundance of particular species could be that those individuals are able to compete more effectively for resources that are shared among multiple species. We summarize these three different

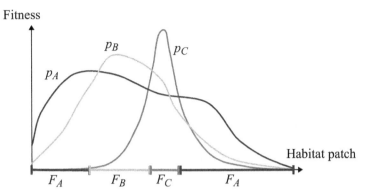

Figure 3.4: The "fitness" associated with three different phenotypes, p_A, p_B, p_C, is plotted against variations in the habitat patch. Phenotype A is managing to cope reasonably well across all patches, making it a generalist, while Phenotype C is much more of a specialist, and requires a very particular environment to be successful. Phenotype B is somewhere in between. The associated fitness sets are shown on the x-axis, i.e., the environmental conditions where that particular phenotype is best suited over the other phenotypes. The fitness set associated with Phenotype A, F_A, is quite large, while, as expected, F_C is significantly smaller.

factors that combine together to determine how well a particular species is able to populate and cover an environmental patch [81, 357]:

1. The availability of resources;
2. The ability to exploit the resources; and
3. Interactions among species that determine their relative successes at utilizing the resources.

These observations bring us to the notion of an *ecological niche*, as introduced in [192], where the niche concept was envisioned as a multidimensional "hypervolume," with the different dimensions corresponding to environmental conditions and resources. This construction is supposed to encompass and express the full relationship between an individual and *all* aspects of its environment. The niche hypervolume includes facets such as feeding opportunities, tolerance to physical conditions (e.g., temperature, insolation, acridity, humidity, salinity, and so on), and avoidance of predation. The ability to avoid being eaten by a predator, for example, would in turn be encoded through other sets of environmental factors, such as the availability of so-called escape spaces. These, in turn, include parameters such as the color and pattern of the backgrounds (which is useful for prey that try to blend in with their surroundings to avoid detection—known as cryptic prey), the availability of hiding places and other areas of refuge, and the structure of the ground vegetation, which is critical for animals that rely on fleeing as a mode of escaping predation [192, 308, 357].

Without becoming overly technical, for the purpose of the developments in this book, it is enough to think of the ecological niche as a particularly well-suited set of environmental conditions for an organism (or robot), when it comes to surviving successfully and productively. But, as we will see in subsequent sections, stand-alone, environmental factors are not enough by themselves in order to delineate organisms and their environments. One also must take into account the way the organisms interact with other organisms, as this too has significant impact on their lifestyles. Robots deployed in natural environments, performing tasks such as monitoring and exploration, are expected to collaborate to achieve their missions. As a result, interactions also play a key role in this setting. And, as already seen in the previous chapter, biology has a lot to teach us about how to think about collaborative teams of robots.

3.2   Interactions

Key to comprehending animal behaviors, and how the animals spend their time and energy, is the way they interact with each other, both within the same species and with members of other species within the same community. This section summarizes some of the different forms these interactions can take, while being on the lookout for potential connections to robotics. It starts with a high-level discussion about populations, followed by a more detailed description of different forms of interactions, such as competition for resources, predator-prey and parasite-host relationships, and mutualism.

3.2.1   Fecundity and Survival

Individuals of the same species in a given, prescribed environmental domain form a population, where variations in the suitability of the habitat lead to changes in the total population, the distribution of individuals across the domain, and even the movement of individuals within the population. The way these habitat variations impact the population size is through the already established and rather straightforward relationship: "changes in population equal births minus deaths." Or, using different terminology, *fecundity*, i.e., the ability to produce an abundance of offspring, and *survival* drive the evolution of the population size. From a robotics point of view, survival is by now such a familiar theme that it even got its own chapter (Chapter 2). Fecundity, on the other hand, is not so clear when designing robots and robotic strategies. However, one can unpack what conditions lead to fecundity and understand those in terms of successful robotic endeavors, without having to add in the somewhat cumbersome notion of robotic "offspring."

An abundance of food (or other sources of energy), and an absence of enemies lead to increased survival and, as such, to an increasing population. But, these factors also lead to increased fecundity, since more nutritious food supports more and healthier babies. Similarly, if time can be spent caring for newborn offspring rather than worrying about being attacked, the fecundity increases as well. In contrast, limitations to the production of offspring can be due to such factors as a shortage of resources, including food sources or places to nest, or, equally problematic, an inaccessibility of these resources relative to the animal's capacity to utilize them [192, 308, 357]. These principles certainly make sense also for robots that need energy to function, and accessible areas to operate in, in order to fulfill their missions. As such, if one squints a bit, *the broad factors that increase animal fecundity can also improve robot effectiveness.*

Resources and predators are not the only limiting determinants to population growth. As this section focuses on interactions, it should be noted that population numbers are highly sensitive to, and regulated by, the effects of density-dependent factors [308, 357]. If there are too many individuals in a small area, the mortality rates go up, and fecundity is reduced. As such, interactions in terms of collaboration, crowding, territoriality, and competition must be taken into account—for robots and animals alike. Taken together, what these observations mean is that the population size is certainly not arbitrary. Rather, it is a well-balanced function of a number of contributing factors.

The notion that there is a particular number of individuals best suited for a given environmental condition or task has been investigated also in the context of robotics, e.g., [68, 276]. For example, one can ask how many transport robots are needed to successfully move an object between locations in a collaborative manner. Mathematically speaking, this is highly related to the question of how many pursuers (transporters) are needed to successfully capture a prey (object to be transported). This connection is made in [99], where the collective transport problem is linked to the way lions hunt, using the

Figure 3.5: Top: One "lion" robot is not able to catch the "gazelle" robot. Bottom: Three lions can indeed capture the gazelle [99].

characteristic "catcher's mitt" formation [91, 136]. In Figure 3.5, it is illustrated how one robotic "lion" is not sufficient to catch a robotic "gazelle," while three lions are found to be the smallest number needed to guarantee capture. If, in addition, the hunters have to share their prey, having more hunters participate means less food for everyone. As such, the optimal number (three, in this case) is obtained. Other manifestations of this idea that there is an optimal number of robots needed for a particular task can be found in different contexts and application domains, such as exploration and coverage tasks, collective transport, and boundary protection, e.g., [142, 175, 180].

3.2.2 Competition

One of the more stark aspects of life in the animal kingdom is that it is defined by a relentless competition for resources. This observation has been an integral part of ecology already from the very beginning of its existence as a scientific discipline. Charles Darwin wrote in "The Origin of Species" [105] that "the prodigious number of plants which in our garden can perfectly well endure our climate . . . will never become naturalised, for they cannot *compete* with our native plants" (emphasis added).

Any time multiple consumers are trying to utilize the same constrained resource, they, almost by definition, have to compete—what is consumed by one animal is no longer available to other animals. In ecology, the distinction is made between intraspecific (within the same species) and interspecific (between individuals belonging to different species) competition [357]. These two types of competition generally have very different effects on the populations; intraspecific competition has a non-catastrophic, regulatory effect on the population size through birth and death rates that drive the population towards an environmentally supported equilibrium, e.g., as per the logistic equation in Equation 3.1. Interspecific competition, on the other hand, can have similarly regulating effects if the competing species are relatively well-balanced, but it can also have a catastrophic impact on one of the competing populations and may even, under particularly unfortunate and sinister conditions, lead to complete extinction.

Regardless of the variety, competition typically occurs as the result of a small set of drivers [357]:

- *Consumptive Competition:* Multiple individuals are trying to directly utilize the same, renewable resource;
- *Preemptive Competition:* An individual is occupying a space, thereby excluding others from that space, in order to be better placed to utilize the available resources;
- *Overgrowth Competition:* An individual (typically of the plant variety) grows over another, thereby getting an advantage in terms of access to light, nutrients, or other resources;
- *Chemical Competition:* An individual produces toxins that act through an environment as a way of removing or preventing others from utilizing the environment productively;
- *Territorial Competition:* An individual defends space for the purpose of securing the exclusive use of the resources in that space; and
- *Encounter Competition:* Individuals engage in transient interactions with each other, which may result in physical harm or theft of food.

Recalling the logistic population model in Equation 3.1, the generalization to two competing populations, while adhering to the same "logistic" form, is the *competitive Lotka-Volterra* model,[5] where the interactions between species are explicitly included. If N_1 is the size of Population 1, and N_2 of Population 2, the competitive Lotka-Volterra model states that

---

[5]The *competitive* Lotka-Volterra model differs slightly from the standard Lotka-Volterra model, where a direct predator-prey relationship is captured rather than the indirect competition of two species through their resource usage [163].

$$\dot{N}_1 = r_1 N_1 \left(1 - \frac{N_1 + \alpha_{12} N_2}{K_1} \right)$$

$$\dot{N}_2 = r_2 N_2 \left(1 - \frac{N_2 + \alpha_{21} N_1}{K_2} \right).$$

(3.4)

Here, α_{ij} encodes the effect that Species j has on Species i. In general, there is no reason to believe that these interaction coefficients should be symmetric, since competitions tend to have winners and losers. However, as the model focuses explicitly on competitive effects, the interactions are assumed to be harmful, which in turn translates to strictly positive coefficient values. Similarly, the growth rates and carrying capacities are also not expected to be symmetric, since each species fits the environmental conditions to a varying degree.

Figure 3.6 illustrates how the competitive Lotka-Volterra model behaves when two species start out with the same population sizes at $N_1 = N_2 = 25$, with the carrying capacities $K_1 = 80$ and $K_2 = 110$, i.e., the environment can support more individuals belonging to Species 2 than to Species 1, and with the growth rates $r_1 = 0.2$ and $r_2 = 0.4$, i.e., Species 2 grows at an inherent faster rate than Species 1. The figure covers the following three different, competitive scenarios:

- *Scenario 1 (Top Right):* $\alpha_{12} = 0.2$, $\alpha_{21} = 0.2$, i.e., both species have a similar, negative influence on each other. As a result, the equilibrium is one where Species 2, which has a higher inherent growth rate and carrying capacity, ends up with a larger population than Species 1. But, due to the competition, neither species reaches its carrying capacity.
- *Scenario 2 (Bottom Left):* $\alpha_{12} = 0.2$, $\alpha_{21} = 0.8$. Here, Species 1 is more harmful to Species 2 than the other way around. The resulting equilibrium has Species 1 with a larger population than Species 2, despite the latter's higher growth rate and carrying capacity.
- *Scenario 3 (Bottom Right):* $\alpha_{12} = 0.8$, $\alpha_{21} = 0.2$. The result from Species 2 growing faster, having a higher carrying capacity, and competing more effectively for the resources, is that Species 2 reaches its carrying capacity, while Species 1 goes extinct.

The two-species model can be generalized to capture competitive interactions among M species in the following, analogous manner,

$$\dot{N}_i = r_i N_i \left(1 - \frac{\sum_{j=1}^{M} \alpha_{ij} N_j}{K_i} \right), \quad i = 1, \ldots, M,$$

(3.5)

where the self-interaction coefficients are given by $\alpha_{ii} = 1$.

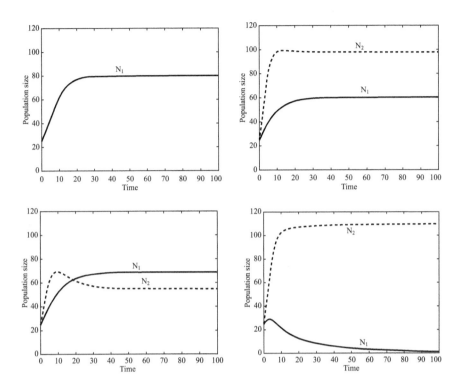

Figure 3.6: A single species' population, evolving according to the logistic equation in Equation 3.1 (under the parameters describing Species 1), is shown together with three different scenarios, where the competitive Lotka-Volterra model (Equation 3.4) predicts qualitatively different outcomes from the interspecies competition, between two populations, as a function of the interaction coefficients.

What the three competitive scenarios in Figure 3.6 call out is the well-established, ecological fact that not only does the inherent environmental conditions determine how well a particular species is doing, but also the way it interacts and competes with other species dictates its success (or lack thereof). In fact, if the interspecific factors, as captured by the α_{ij}-terms, are relatively large and dominate the intraspecific factors (the r_i and K_i-terms) then, according to the competitive Lotka-Volterra model, this will inevitably lead to the extinction of at least one of the species [53, 54].

Another way of arriving at the unfortunate situation where one species is completely eliminated is through the so-called *competitive exclusion principle*

[179], also known as Gause's Law[6] [339]. This principle states that when two species are competing for the same constrained resources, both species cannot coexist. Whenever one of the species has even the slightest advantage, it will eventually and completely dominate. The result from this superiority is either extinction of the weaker population or, more commonly, a shift towards a different ecological nice.

Although there may not be an immediate analogy for the types of competition captured by Lotka-Volterra—intraspecific as well as interspecific—in long-duration autonomy, what this somewhat cautionary tale of competitive exclusion tells us is that even for robots deployed in an environment over long time-scales, the availability of resources does not tell the full story. The competition for (or, at least, the necessity of sharing) these resources matters as well. This is something that must be captured and considered by a robot ecology design paradigm for long-duration autonomy.

3.2.3   Predators and Parasites

Continuing down the path of characterizing how competition, as a particular interaction modality, imposes constraints on the sustainable population sizes, we now shift focus to two particularly pertinent forms of population-size regulation mechanisms, namely *predation* and *parasitism* [2, 144]. Consumers typically fall into two[7] distinct camps, depending on their resource consumption strategies, namely predators, which catch other individuals and, bluntly put, eat them, and parasites that consume a host that keeps living through the process.[8] In other words, predators remove their prey from the population, while parasites do not—at least not directly. A host suffering from a parasite may have its fecundity reduced or experience an increased probability of dying from other causes. But, the host is not directly removed from the resource population through the act of parasitism.

---

[6]Gause's Law is named after the ecologist Georgy Gause, even though variations on the competitive exclusion principle were formulated already in Darwin's Theory of Natural Selection [105, 157].

[7]A third category is made up of *parasitoids*, which include species of wasps and flies, whose larvae consume the living host, eventually leading to the death of the host [419].

[8]What about herbivores that eat plants? They are neither predator nor parasite, right? Actually, they can be either. They can be a plant predator in that they consume the whole plant, e.g., sparrows that eat seeds, thereby killing off the entire embryo of the plant, or parasites in that they consume living plant tissue but do not kill their victims, such as deer browsing on shrubs [427].

Even though it is fascinating and a bit horrifying to imagine predator robots, the primary, relevant aspect of this connection is the mobility strategies that consumers and resources (predators and prey, or parasites and hosts) employ in order to eat/not get eaten. The reason these strategies are relevant is that a number of robotic tasks involve security and defense-related activities, such as protecting an area from intruders—thereby having the robots move like predators—or monitoring an environment, like the ocean floor, without getting caught by malicious or just clueless actors, such as passing ships—thereby moving like prey. In fact, prey have a few possible tools at their disposal to avoid predation, such as *hiding*, *fleeing*, or *fighting*. Predators, in turn, typically have to be significantly larger than their prey in order to effectively catch and subdue them. However, a few species, such as wolves, hyenas, lions, and army ants hunt cooperatively in order to be able to take down larger prey, e.g., [173].

Cooperative hunting manifests itself in different ways, but the basic calculus is that it is more efficient for a group of predators to work together in order to maximize the expected caloric intake. Depending on what cooperative strategy is used, and the characteristics of the prey, many different types of *geometrically* defined, collaborative hunting strategies exist in nature. This is quite promising since, as seen in Chapter 2, when designing multi-robot behaviors for teams of robots, geometrically defined costs and shapes oftentimes serve as the starting point for the design process. Moreover, geometries provide natural modalities for human users to interact with robot teams using instructions like "form a triangle," "spread out across an area," or "stay close to each other," e.g., [118]. To this end, we are particularly interested in this idea of geometrically defined hunting strategies, and in what nature has to say about the geometry of the predator-prey interaction.

One of the drivers behind the geometry of the hunting strategy is the question of who gets to eat. Just because a dolphin has caught a mackerel, the rest of its pod-mates do not necessarily get to eat. But, if a lion catches a zebra, the whole pride gets to feast [178]. Clearly, the calculus is different in these two cases, where the lions should maximize their chances of capturing a prey across the entire team, while the dolphins may want to maximize the average (or minimum) amount of fish caught by any individual dolphin. To further elucidate this point, bottlenose dolphins, *Tursiops truncatus*, and African lions, *Panthera leo*, both charge towards an aggregation of prey (a school of fish or a herd of zebras) by forming a coordinated predator front [136, 343]. They arrange themselves in specific formations—for example, the lion front is a U-shaped "catcher's mitt" formation [136]—to create a front that moves together in unison towards the collection of prey. In contrast to the catcher's mitt, porpoises, i.e., the family of dolphins and whales, employ different and oftentimes quite elaborate strategies when hunting [259, 343, 372]. Bottlenose dolphins have been known to forage for fish using well-coordinated formations, with suggestive names like the "wall method" or the "horizontal carousel." Under the former, a group of dolphins drive their prey towards a barrier, and feed off the returning, rebounding fish. Under the latter, the dolphins first encircle a group of fish and subsequently tighten this encirclement to restrict the movement of their prey. At some point throughout this process, they cease the encirclement and instead charge through the school of fish, one at a time, or all at once [177, 259, 343, 372].

Given this rich variety of strategies and geometrically defined hunting procedures, one can thus ask questions with clear and direct relevance to robotics about what the optimal shapes are for the predator fronts in the context of foraging animals (or multi-robot systems). In [178], this question was pursued by modeling the shape of the predator front as a curve, and the total energy intake, i.e. the total amount of prey swept by the front, was maximized by modifying the shape. One application for this type of inquiry is robotic cleanup of oil spills, as shown in Figure 3.7 (Top Left), where a group of robots drive a flexible suction boom towards a spill site, and where these biologically based considerations can help optimize the shape of the boom to clean up the oil spill.

The prey motion is modeled as a reaction-diffusion process in [178], with the predator front charging through the area containing the prey. The diffusive component corresponds to the unperturbed movement of the prey, while the reactive part captures how the prey responds to the predators, describing how the prey animals flee in the presence of the predators. The effect of optimizing the predator front is shown in Figure 3.7, where the initial prey density (Top Right) is given by a tall and skinny rectangular shape. The curves displayed in the bottom figures are the optimal predator fronts, given increasing reactive gains in the reaction-diffusion process. As expected, when the prey is less scared of the predators (Bottom Left) the shape is less elaborate, as a simple sweep through the prey distribution results in a highly successful hunt—like shooting fish in a barrel. If the prey is more skittish (Bottom Right), the predators must quickly get to the center of the prey density. Hence the more pronounced peak in the middle of the predator front.[9]

---

[9]Other geometrically defined, bio-inspired foraging strategies, based on ants and bees can, for example, be found in [70, 115, 190, 222, 231].

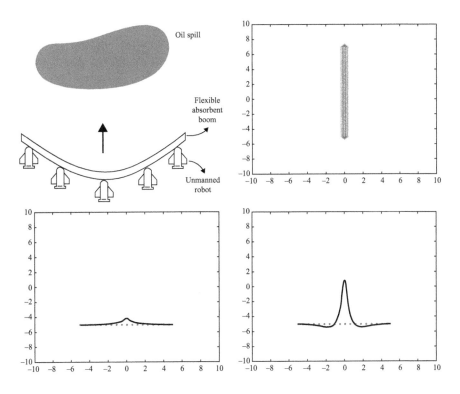

Figure 3.7: Top Left: Unmanned vehicles are driving a flexible, absorbent boom towards an oil spill. The shape of the boom can be optimized based on predator-prey interaction models. Based on the findings in [178], the shape of such a predator front is optimized as it sweeps through a prey density. Top Right: Initial prey density. Bottom Left: Optimal predator front when the prey's motion is primarily diffusive. Bottom Right: Optimal front when the prey is trying to evade the predators according to a reaction-diffusion process, with a high-gain reactive component. Source: Fig. 1(a) and Fig. 4 in Haque et al., 2014.

The findings in [178] show that the shape of the optimal predator front depends on how rapidly the prey responds to the predators, i.e., once again the time-scales of the interactions between individuals, as well as with the environment dynamics, play a central role in the viability of the behaviors. The importance of the temporal alignment between behaviors and environment was in fact one of the cornerstones of robot survival in Chapter 2. As such, it does not come as a surprise that it crops up also in the context of predator-prey interactions. To consume prey, the predators first have to catch it.

It should be noted, though, that not all interactions—robotic or otherwise—are harmful and confrontational. This section concludes with a a brief discussion of some of the many other types of robotics-relevant interaction modalities encountered in nature, lumped together under the umbrella term of "social behaviors."

3.2.4 Social Behaviors

In terms of animals' fitness, their strategies and behaviors can be more or less well-suited to particular environments. One way of increasing an animal's fitness is by teaming up and cooperating with other individuals. The resulting interactions are certainly less confrontational and harmful than the previously discussed predator-prey or parasite-host relationships. Instead, they form the basis for a rich class of social behaviors that animals exhibit as a way of making their existences more successful in terms of survival and fecundity [107].

Cooperation, broadly speaking, means that individuals work together towards common and mutual benefits, as opposed to the competitive situation, where individuals are actively opposing each other in order to realize more selfish goals. A particular poignant form of cooperation is *altruism*, where an individual is even willing to sacrifice its own well-being or potential benefits in order to promote other individuals. These altruistic behaviors (and, conversely, spiteful[10] behaviors) are related to the "kin selection" process, which encompasses a number of scenarios where an individual acts in order to benefit the reproductive success of relatives [44]. The so-called Hamilton's Rule states that an altruistic act is beneficial and should be performed if

$$rB > C, \tag{3.6}$$

where r is the "relatedness" between the individual committing the altruistic act and the beneficiary of the act, B is the benefit to the recipient, and C is the cost to the "altruist" [44].

Apart from the absolute impossibility of associating units with these quantities (or somehow measuring them), it is an instructive equation in that it does tell us that an animal is more likely to perform a costly action if the beneficiary of this act is a close relative. And, if we replace "reproductive success"

---

[10]A spiteful behavior is usually the opposite of an altruistic behavior in that it is an act that results in harm, and by extension, loss of fitness, to both the spiteful actor and the recipient.

with just "success," and set $r = 1$ (all robots are related), or possibly $r = 1 - \epsilon$, for some $\epsilon > 0$ to avoid overly selfless and meek robots, then we have a crisp and actionable formula for when a robot should help another robot.

But, kin selection is not the only reason why cooperation and social behaviors are strategically wise. In fact, in many species, social behaviors are exhibited not only between mates and progeny, but also among extended families, and even among large groups of unrelated individuals.[11] This grouping is clearly beneficial, as it helps the animals out-survive and out-produce other groups. The social tendencies can lead to well-defined aggregation of individuals to form tight groups, e.g., salamanders in "clumped" distributions, schools of fish, or flocks of birds. In fact, cooperation and social behaviors within species encompass a number of different facets, such as cooperative breeding (e.g., among the notoriously promiscuous capuchin monkeys) or foraging (e.g., among wolves) [61]. Cooperation can also help with defensive strategies. For example, a type of bee—the *Tetragonula carbonaria*—organizes itself in a so-called fighting swarm for the purpose of defending itself against attackers [433].

As an example of how cooperative behaviors can manifest themselves in elaborate social structures, let us return to the bottlenose dolphins, who lead highly social existences in pods. Male bottlenose dolphins form two levels of social groupings, namely first-order and second-order alliances [87, 90, 88, 89, 176]. These alliances are built to increase a male's chances of mating. The first-order alliance consists of a pair or triplet of male dolphins, and dolphins are almost always seen together with their first-order alliance members. More formally, they share a high "association coefficient" [87], which is an indicator of how often two individual dolphins are seen together. Males of a first-level alliance herd female dolphins by swimming in a coordinated manner, and this way they inherently constrict the movement of the elusive females. However, a first-order alliance is incapable of stealing a female dolphin already being herded by another alliance [88]. In such a scenario, the dolphins coordinate with each other to build higher-level, second-order alliances.

The high-level takeaway from this section is the observation that cooperation can be highly beneficial for the purpose of more effective utilization of resources. This could be due to a more complete coverage of an area, a tighter

---

[11] Sometimes cooperative "agreements" are even reached between members of different species. *Mutualism* expounds this process, encompassing interactions among different species, where each species receives a net benefit from the cooperation [61].

security perimeter, a more effective hunting endeavor, or a tag-teaming ap-
proach to tasks that require a persistent performance. These are all reasons
that resonate well with the robot ecology paradigm, where cooperation among
robots is not only potentially beneficial, but oftentimes required.

3.3   Ecologically Inspired Constraints

"Even the most slowly reproducing species would cover the earth in a
short time if its population growth were unconstrained" [357].

Alas, this suggestive quote can never be fully put to the test and validated
as no organism ever has found itself in the enviable position of having
access to completely unlimited resources. In fact, as resources are always
limited, compromise and the balancing of constraints are central to all life
forms.

As discussed repeatedly in different contexts already, such resource constraints are largely functions of the environment in which the organism
resides. The environmental, ecological constraints serve not only as a brake
on a species' population growth, but also give rise to organismal traits that are
uniquely adapted to a particular habitat [7, 102]. These traits manifest themselves through two primary components of an individual's fitness, namely its
ability to *survive* and to *reproduce* successfully, e.g., [7]. In other words, organisms must first meet the demands of the environment to survive, and then
successfully reproduce to pass on their genotypic and phenotypic traits—
surviving is a prerequisite to thriving. Survival, in turn, involves navigating
energetic landscapes, meeting nutritional needs, and avoiding predation, while
reproduction involves strategies and behaviors for successfully finding a mate
and producing viable offspring.

The takeaway from this discussion is that survival in nature is a direct
function of the organism's ability to satisfy ecological constraints. Luckily for
us, one of the key design principles for long-duration autonomy, articulated
already in the two introductory chapters, is that robot survival is also naturally
expressed through *constraints* rather than costs. What this means is that no
matter what the robot might be up to, it should always satisfy the survival
constraint, expressed vaguely yet illustratively as

$$
\begin{aligned}
&\text{minimize } \mathscr{C}_{task} \\
&\text{subject to } x \in \mathscr{G}_{survive},
\end{aligned}
\tag{3.7}
$$

where x is the state of the robot, as per Equation 1.1. What we will do, for the remainder of this chapter, is elaborate on what this constraint might entail, once one embraces the ecological view of robots and their environments, by translating the unearthed ecological constraints to a robotics-relevant language.

The translation from ecology to robotics in general, and robot survival in particular, is not that much of a stretch in that, as seen repeatedly, richness of behavior in the natural world is largely driven by constraints imposed on the individual by the limited environmental resources, and by intra- and inter-specific competition. The notions of "survival" and "constraints" are thus central to both fields. In other words, the *mutual intelligibility*, to borrow a term from linguistics, between the languages of ecology and long-duration autonomy is quite high.

3.3.1 Ideal Free Distributions

Foraging provides a rather direct animal-to-robot link in that animals and robots alike have to acquire the necessary energy in order to survive. Foraging, in turn, has defining implications for an organism's space use and selection of particular habitat types. An organism must forage across a sufficient area to support its energetic needs. Additionally, if critical resources (e.g., particular diet items or water sources) are only available at select locations, the organism is "spatially anchored" [378] and is forced to return to those locations. The corresponding constraint states that the animal should always have enough energy to return to the spatially anchored feeding locations.

Another set of behaviors that relate to the use and defense of space has to do with *territoriality*. A territory is defined as an exclusive area belonging to an individual or group of individuals within a particular species [154, 308], which already sounds eerily similar to the notion of a *region of dominance*, as defined by the Voronoi cells associated with individual robots in the coverage control problem from the previous chapter (Equation 2.18). Territorial behaviors, in turn, impose constraints on the population density, sometimes with highly confrontational and lethal side-effects. For example, the grain beetles, *Rhizopertha*, are known to fight to the death over a single kernel of wheat [100]. But for most animals, no single "meal" is worth an all-out, to-the-death battle. Instead, the defense of resources is done by defending the territory against intruders. Holding territory offers a number of clear advantages. Those include the exclusive use of resources, reduced predation and disease owing to greater spacing from neighbors, and predator escape facilitated by familiarity with an area [378]. Territoriality can also provide exclusivity in breeding [357]. (Certain animals, like some species of birds, only establish and defend territory during mating season.)

Regardless of why an animal is devoted to a particular geographic area, the ideal free distribution dictates that the animals will be spaced out over the environmental patch in such a way that they all have access to an equal, realized set of resources [330]. To that end, let us start with the ideal free distribution as a way to untangle what territoriality and space usage might look like in a robot control setting. As before, assume that the domain that the animals (or robots) inhabit is given by $\mathcal{D}$, and that $\varphi(p) \in \mathbb{R}_+$ describes the resource density at location $p \in \mathcal{D}$. The total amount of resources available in the environment is thus given by

$$R_{\mathcal{D}} = \int_{\mathcal{D}} \varphi(p) dp. \tag{3.8}$$

Now, if we let $T_i \subset \mathcal{D}$ be the "territory" associated with Individual i (Animal or Robot i), then the total amount of resources available to that individual is

$$R_{T_i} = \int_{T_i} \varphi(p) dp. \tag{3.9}$$

The constraint associated with an ideal free distribution would accordingly state that the individual "resource integrals" be equal[12], i.e.,

$$R_{T_i} = R_{T_j}, \ \forall i \neq j. \tag{3.10}$$

Alternatively, as seen from the logistic equation in Equation 3.1, the carrying capacity, K, is the population size that can be supported by the environment at steady state. By normalizing the total amount of available resources in Equation 3.8 over K, the individual resource constraint could instead be formulated as

$$R_{T_i} \geq \frac{R_{\mathcal{D}}}{K}, \ \forall i. \tag{3.11}$$

But, just because resources are available in an area, it does not follow that an individual can utilize all resources equally well—shorter animals may not

---

[12]Of course, insisting on these quantities being exactly equal is probably too strong. A bound on the allowable differences would make for a more realistic and achievable constraint.

be able to reach as much fruit as taller animals. If we let x_i be the state of Animal/Robot i, then we can let $z_i(x_i, p) \in [0, 1]$ describe how well the resource can be utilized by this individual, with $z_i = 1$ meaning perfect resource utilization, and $z_i = 0$ signifying a complete lack of effective access to the resource in question. Accordingly, the utilization-weighted resource integral would change to

$$R_{T_i} = \int_{T_i} z_i(x_i, p)\varphi(p)dp, \tag{3.12}$$

with the resource constraint, $R_{T_i} \geq c_i$ where, $c_i > 0$ is the amount of effectively utilizable resources needed for Individual i to survive.

One can imagine a number of variations on this theme, such as the addition of multiple species, integrals defined across populations instead of individuals, and the inclusion of more than one type of resource [330]. This is analogous to what is done in [364], where robots with heterogeneous and qualitatively different sensing capabilities are supposed to cover an area, and where some sensors can detect certain environmental effects but not others. Additionally, one can relax the insistence on a crisply defined region assigned to particular individuals, and instead allow for the regions to bleed into each other. This is not without precedence in nature. For example, the dragonfly (*Leucorrhinia rubicunda*) switches from strong "site fidelity" (tightly controlled territories) at low densities, to much more loose and overlapping feeding areas at high densities [323].

Regardless of which variation on the resource integral one goes with, it is clear that one key, constituent part in the survival constraint, $x \in \mathscr{G}_{survive}$, is that $x \in \mathscr{G}_{resources}$, i.e., the organism must have access to enough resources for survival to be possible. Or, put another way, surviving equals having enough resources *and* possibly satisfying other constraints as well. The set-theoretic operation corresponding to logical *conjunction* ("and") is intersection, i.e., we can begin to add structure to the survival constraint,

$$\mathscr{G}_{survive} = \mathscr{G}_{resources} \cap \mathscr{G}_{other}, \tag{3.13}$$

where $\mathscr{G}_{other}$ is a set of additional, yet to-be-defined, components of the survival constraint.

3.3.2 Competitive and Cooperative Interactions

One of the rationales behind pursuing long-duration deployments is that a persistent presence in an environment provides access to measurements and other capabilities that would otherwise not be possible. Examples of such capabilities include environmental monitoring and surveillance, area protection, exploration, and search-and-rescue. What all of these application scenarios have in common is that they are more effectively executed if more than a single robot is partaking in the mission. As such, robot-robot interactions matter. In nature, populations interact primarily through different types of "feeding" relationships, e.g., [102], with the predator-prey relationship being an example of a direct such relationship, while consumers sharing common resources are indirectly interacting through the resource consumption. But, as seen in Chapter 2, the *geometric* characteristics of multi-robot configurations are oftentimes what defines the robot interactions. As a consequence, rather than overfocusing on predation or feeding, we instead approach the topic of robot-robot interactions from the vantage point of geometric constraints, pertaining to robot survival. The expectation is, however, that such geometrically defined constraints could be used (at least partially) to express the motivating, ecological interactions.

Going back to Reynolds' Boids model in Chapter 2, not getting eaten by a predator could be mapped onto a separation constraint, i.e., no robots should get too close to each other for the purpose of avoiding collisions. If x_i and x_j are the positions of Individuals (Animals/Robots) i and j, then this constraint would dictate that

$$\|x_i - x_j\| \geq D_s, \ \forall i \neq j, \tag{3.14}$$

where D_s is a given safety distance.

Not being too close to other individuals does not tell the full story, however. For the team of robots to be able to function correctly, they need to be able to share information, and to learn from and coordinate their behaviors with each other [68, 276]. This is only possible if the underlying network topology that encodes the information exchange among robots—may it be sensing- or communications-based—is *connected*. What this entails is that there must exist, at the very least, a path through the network between any two robots in the team [162].

Connectivity is a global, team-level property, rather than a pair-wise property, and, as seen in [162, 276], an undirected network is connected if and only if the so-called *algebraic connectivity*, λ_2, of the network is positive, i.e.,

$$\lambda_2(x) \geq 0, \tag{3.15}$$

where x is the composite, joint state of all robots in the team. Without going into detail about how the algebraic connectivity is computed, or even how the robot network is defined, for the purpose of this exposition it suffices to observe that it is in fact a geometric constraint,[13] just like Equation 3.14. As such, it fits nicely into the broader narrative around interactions as constraints.

But, unfortunately, just staying connected is typically not enough. Let the state-dependent interaction graph that encodes the way the robots interact be given by $G(x)$.[14] Depending on what one wants the robot team to accomplish, there are additional, network-topological constraints that must be satisfied by $G(x)$. For example, in order for the robots to execute the cyclic pursuit behavior in Equation 2.7, $G(x)$, must contain a spanning cycle, C_N, [261]. The corresponding interaction constraint thus becomes $C_N \subseteq G(x)$.

If, instead, what is attempted by the robots is the assembly of a particular geometric shape, the success of this endeavor depends in a rather intricate way on a connection between the robots' sensing modalities and the network topology. If the robots are equipped with range sensors, i.e., they can measure distances to nearby robots, then the graph must be (infinitesimally) *rigid* [15]. If, instead, they are equipped with monocular cameras and can, as a result, only measure the relative angles between robots, the graph must be (infinitesimally) *shape similar* [67]. Rather than covering this topic in detail, the main point to be made here is that the interactions, yet again, impose constraints on the feasible behaviors of the robot team.

Another manifestation of these interaction-based constraints can be found by returning to the coverage control problem, where coverage behaviors can be designed as gradient descent flows to locational costs, akin to the recently introduced resource-integrals. But this construction only works if individuals can exchange information with those with whom they share a boundary in the Voronoi tessellation. The graph structure that encodes this relationship is known as a Delaunay graph, G_D, i.e., we need the constraint $G_D \subseteq G(x)$ to hold for the gradient descent flow to achieve proper coverage of the area [68].

All of these multi-robot scenarios and their required, supporting network topologies impose stronger constraints on the interaction graphs than just connectivity, i.e., in all of those cases, connectivity is a necessary but not sufficient

---

[13] This holds as long as the information-exchange network is geometrically defined, e.g., robots are adjacent in the network if they are "close enough" to each other, given some suitable notion of proximity among robots, [275].

[14] Such multi-robot interaction networks will be covered in detail in Chapters 6 and 7.

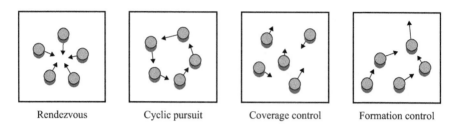

| Rendezvous | Cyclic pursuit | Coverage control | Formation control |

Figure 3.8: Four multi-robot behaviors that require different interaction constraints (network topologies) to be satisfied to function properly: Rendezvous (requires connectivity); Cyclic pursuit (cycle graph); Coverage control (Delaunay graph); and Formation control (infinitesimally rigid graph).

condition for a graph to be cyclic/rigid/Delaunay, etc.[15] These representative examples of multi-robot scenarios are illustrated in Figure 3.8, and their associated interaction constraints are summarized below.

Multi-Robot Behavior	Network Constraint
Rendezvous	Connectivity $[\lambda_2(G(x)) \geq 0]$
Cyclic Pursuit	Cycle Graph $[C_N \subseteq G(x)]$
Coverage Control	Delaunay Graph $[G_D \subseteq G(x)]$
Formation Control	Graph Rigidity/Shape-Similarity

Gathering these observations together, we can absorb all of the different types of interaction constraints into one grand, unified constraint, $\mathcal{G}_{interact}$, that should contain safety in terms of enforcing sufficient separation between individuals, connectivity in order to ensure that effective information-sharing is possible, and possibly additional topology constraints, depending on the particulars of the task being pursued. In other words, this unified interaction constraint would take the form

$$\mathcal{G}_{interact} = \mathcal{G}_{separation} \cap \mathcal{G}_{connect} \cap \mathcal{G}_{topology}. \tag{3.16}$$

The sought-after survival constraint can thus be further refined as

$$\mathcal{G}_{survive} = \mathcal{G}_{resources} \cap \mathcal{G}_{interact} \cap \mathcal{G}_{other}, \tag{3.17}$$

---

[15]In order to solve the rendezvous problem using the consensus equation in Equation 2.6, connectivity is in fact both necessary and sufficient [276].

where some additional work is needed to finalize the missing $\mathcal{G}_{other}$ ingredient.

3.3.3 Thermoregulation and Task Persistification

One aspect of life in the natural world that consumes significant effort and energy is the maintenance of internal functions; in particular regulation to a constant, internal temperature. As previously discussed, this activity comes in two distinct varieties, namely physiological thermoregulation, through internal means, and behavioral thermoregulation, through external means, such as seeking shade to keep cool, or huddling together for warmth. Phrased as a constraint, if $\mathcal{T}$ is the internal temperature, the thermoregulation constraint would take the form

$$\mathcal{T} \in [\mathcal{T}_{min}, \mathcal{T}_{max}], \tag{3.18}$$

where the tightness of the interval is a function of the particular animal's tolerance to internal temperature variations. To complicate matters, $\mathcal{T}$ is a function of both the animal's state (internal and external) as well as an abstract, environmental state. Chapter 5 will focus extensively on how these quantities should be modeled and coupled. However, for a robot, the maintenance of a narrowly defined internal temperature is typically not all that crucial. Instead, we could take the thermoregulation idea and map it onto another key, internal quantity that must be kept within reasonable levels for a robot (or animal) to function effectively, namely its energy cache.

Let E be the energy level of a robot's battery. Long-duration autonomy means that this quantity must remain in the interval $[E_{min}, E_{max}]$, which, in turn, can be thought of (if one squints a bit) as "robot homeostasis." To this end, if $\mathcal{G}_{thermo}$ is the set that captures the thermoregulation constraint, we can let $\mathcal{G}_{homeostasis}$ be its generalization, as homeostasis refers more broadly to the ability to maintain constant (or almost constant) internal conditions.

An added bonus with phrasing energy-management as a homeostasis constraint,

$$E \in [E_{min}, E_{max}], \tag{3.19}$$

is that it points directly to the need to *persistify* robotic tasks, i.e., to the need to be able to modify whatever the robots might be doing in such a way that the energy constraint is always satisfied, no matter what, [302, 304]. In fact, as this constraint plays such an important role when ensuring robot longevity, it will be the topic of Chapter 5, where the notion of "task persistification" will

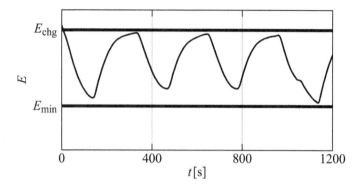

Figure 3.9: A robot's battery levels during a persistent, environmental monitoring experiment. The persitification strategy ensures that the robot's energy level always falls within the prescribed bounds.

be formally investigated as the mechanism by which a robot is guaranteed to be able to carry out its mission indefinitely—beyond any finite number of battery charges/tanks of gas.

To foreshadow the discussion in Chapter 5 slightly, Figure 3.9 shows the battery level associated with a robot that is executing a persistified version of an environmental monitoring task. As can be seen, the energy level does indeed stay within the prescribed bounds as a direct consequence of the enforcement of the robot homeostasis constraint in Equation 3.19.

Purposefully ignoring the (from a robotics vantage point) slightly complicated issues of mating and the drive to produce offspring, the ecological connections associated with the consumption of resources, effective utilization of space, interactions with other individuals, and maintenance of internal conditions, come together to paint a reasonably complete picture of what the ecological survival constraint might look like to a robot. In other words, the thermoregulation/homeostasis constraint constitutes the final piece in the set-theoretic survival puzzle. And, what we require of a robot, acting in accordance with ecological principles, is that it should satisfy the survival constraint, $x \in \mathcal{G}_{survive}$, where the survival set is given by

$$\mathcal{G}_{survive} = \mathcal{G}_{resources} \cap \mathcal{G}_{interact} \cap \mathcal{G}_{homeostasis}. \tag{3.20}$$

3.3.4 Towards Robot Ecology

We have now seen a number of attempts at translating ecological principles and ideas to concepts and constraints that have technical meaning in robotics.

One can think of these attempts as ecologically inspired robotics. But that is not the same thing as "robot ecology," which is what we are ultimately after. To get to that, we have to somewhat reverse the order of the story. In other words, we need to relate robot tasks and considerations to ecological constraints (rather than the other way around, as was done in this chapter). By doing so, we will eventually arrive at a theory of robot ecology, i.e., a formal framework for robots deployed over long time horizons, where the behaviors are tightly coupled to the deployment environments. This framework will have to be firmly grounded in ecological principles, not for ecology's sake, but for the sake of harnessing the power of a field that focuses explicitly on the coupling between organisms and their habitats. To this end, we conclude this chapter with a few of the discussed, ecologically grounded principles, highlighting what the reversal of the story might look like.

Design Principles for Robot Ecology

Maximize the Fitness Set
Any solution that is too carefully crafted and calibrated to particular, narrowly defined environmental conditions is doomed to fail in a long-duration setting, as the world is dynamic, unpredictable, and messy.

Ideal Free Distributions
When robots are deployed in an environment for long periods of time, ready to be recruited and tasked to carry out different missions, they should be arranged so as to provide effective coverage of the area in terms of available sensor measurements, situational awareness, and other capabilities.

Carrying Capacity
Environments impose limits on how many robots they can support in terms of available energy resources, congestion-free navigation, and avoidance of crowding effects. Similarly, not all robots should participate in all missions, as missions typically require a certain number of robots to be executed effectively.

Collaborative Interactions
The driving applications behind long-duration autonomy, such as environmental monitoring and surveillance, area protection, exploration, and search-and-rescue, require robots to collaborate productively.

These collaborative interactions impose constraints on the possible robot behaviors.

Purposeful Expenditure of Energy

Survival is a prerequisite to thriving. Not getting stranded somewhere with completely depleted batteries and no ability to recharge is therefore of paramount importance. This means both that a careful power management strategy is necessary and that whatever tasks the robots may be executing are rendered energetically persistent so that they can be performed indefinitely.

II Constraint-Based Control

4 Constraints and Barriers

As discussed in the past chapter, environmental constraints, phrased in eco-logical terms such as predation, territoriality, foraging, mutualism, and so on, give rise to a rich and sometimes surprising set of behaviors among animals [101, 137, 308, 357, 385]. Additionally, when robots are to be deployed over long time-scales, surviving, i.e, not ending up in a situation from which the robots cannot recover, takes precedence over acting in an optimal fashion, as *survival is a prerequisite to thriving*. In this chapter, we formalize these observations and, in particular, introduce the technical machinery needed to properly manifest constraint-based control design. The key construct towards this end is that of a Control Barrier Function (CBF), and we will build up to-wards establishing how robot survival can be expressed both compactly and elegantly through the use of CBFs.

Recall the vague yet suggestive statement from Chapter 1 that con-tained the essence of constraint-based control design for long-duration autonomy,

$$\text{minimize } \mathscr{C}_{task}$$
$$\text{subject to } x \in \mathscr{G}_{survive}.$$

Here, x is the state of the robot, and the survival constraint has to hold at each instant of time. But, without explicitly calling out the decision variable over which the minimization takes place, this formulation is hopelessly ill-fated. As such, one needs to add the control input, u, to the formulation, i.e., incorporate the way control decisions affect the state of the robot, which is the only way the task described through $\mathscr{C}_{task}$ can be manipulated. To this end, if we encode how well (or, really, how poorly) the task is being performed through the cost $\mathscr{C}_{task}(x, u)$, that is now a function of both the current state, x, and the control input, u, we could let the robot solve the following constrained optimization problem at each instant of time,

$$
\begin{aligned}
&\min_u \mathscr{C}_{task}(x, u) \\
&\text{subject to } x \in \mathscr{G}_{survive}.
\end{aligned} \tag{4.1}
$$

This formulation certainly looks more well-defined than what was discussed in Chapter 1. Unfortunately, it is still mathematical nonsense. To see why this is the case, consider what is actually going on here. The state, x, is the *current* state of the robot, i.e., it is what it is, and it is *not* a free variable that can be directly manipulated. So, we are asking the robot to minimize a function of u (and of x) by a clever selection of u, subject to a constraint that either holds or it does not—*no matter what value the decision variable, u, takes*. This is why this unfortunate formulation, albeit quite suggestive, is nonsense. What is needed is a way to connect u to x. And the way to do so is through the dynamics of the robot, coupled to the technical notion of *forward invariance*.

4.1   Forward Invariance

4.1.1   Collision-Avoidance

To get things started, consider a self-driving car. One conspicuous constraint that this car needs to satisfy is to avoid collisions, i.e., the position of the car/robot should not coincide with the locations of any objects in the environment, including other cars. To this end, assume that the car is driving rapidly towards a wall. The constraint that the wall and the car should not be in the same location is not a particularly useful constraint as there comes a point, as the car is approaching the wall, when a collision is unavoidable even though it has not yet happened. In other words, nothing the car does from then on will prevent the collision from happening. The genuine constraint needs to be to avoid ending up in this situation in the first place, i.e., the car should never find itself in a situation where a collision cannot be avoided. It is not enough to simply consider the position of the car. Instead, the full state of the

Figure 4.1: Two robots on a string are moving towards each other, eventually making a collision inevitable.

car (position, orientation, longitudinal and lateral velocities, and so on) needs to be taken into account, and the dynamics of the vehicle must enter into the discussion.

To see how this works, consider the rather simplified example of two "robots" moving on a line (like two beads on a string), illustrated in Figure 4.1. If they can directly control their velocities, the states of the robots become their positions, $x_i \in \mathbb{R}$, $i = 1, 2$, with the controlled dynamics being

$$\dot{x}_i = u_i, \quad i = 1, 2, \tag{4.2}$$

where u_i is the control input applied to Robot i.

If there is a desired, minimum safety distance, D_s, that should be maintained, the corresponding collision-avoidance constraint becomes $\|x_i - x_j\| \geq D_s$, or, more conveniently phrased,[1]

$$\|x_i - x_j\| - D_s \geq 0. \tag{4.3}$$

This constraint has to hold for all times in order for the two "bead-robots" to be safe.

However, if instead of velocities, what was controlled was accelerations, the state would be two-dimensional, $x_i = [p_i, v_i]^T \in \mathbb{R}^2$, $i = 1, 2$, where p_i is the position and v_i the velocity of Robot i. The controlled dynamics associated with this configuration are

$$\begin{aligned} \dot{p}_i &= v_i \\ \dot{v}_i &= u_i, \quad i = 1, 2. \end{aligned} \tag{4.4}$$

If, additionally, there was a maximum acceleration imposed on the control input, say $\|u_i\| \leq a$, $i = 1, 2$, then "slamming the brakes" would mean that a

---

[1]Even though these are scalar quantities, we use the Euclidean norm, $\| \cdot \|$, rather than the absolute value, $| \cdot |$, as a way of keeping the notation consistent and foreshadowing the later developments in the chapter.

robot with initial conditions $x_i(0) = [p_i(0), v_i(0)]^T$ would come to a complete stop at the location

$$p_i = p_i(0) + \frac{1}{2} \frac{v_i(0)^2}{a} \operatorname{sign}(v_i(0))a,$$

where we have interpreted "slamming the brakes" as using $u_i = -\operatorname{sign}(v_i(0))a$.

Similarly, with two robots, their distance from each other, $\|p_1 - p_2\|$, will stop shrinking under the worst-case scenario (the robots are initially driving towards each other) at a final distance of

$$\|p_1 - p_2\| = -\frac{1}{4} \frac{\|v_1(0) - v_2(0)\|^2}{a} + \|p_1(0) - p_2(0)\|.$$

But, if we changed perspective and instead viewed the initial condition as the current state of the robot, and insisted on the final relative distance (at a stand-still) being greater than the safety distance, D_s, the collision-avoidance constraint would become

$$-\frac{1}{4} \frac{\|v_1 - v_2\|^2}{a} + \|p_1 - p_2\| - D_s \geq 0. \tag{4.5}$$

As a result, we have two rather different looking safety constraints (Equations 4.3 and 4.5, respectively) depending on what the dynamics of the robots are. For this reason, we will keep the discussions general by initially considering robot states, $x \in \mathbb{R}^n$, governed by the control-affine dynamics,[2]

$$\dot{x} = f(x) + g(x)u, \tag{4.6}$$

where $f(\cdot)$ is the drift term (what the robot would do if there were no inputs), $g(\cdot)$ is the controlled dynamics (the effect the actuators have on the robot), and $u \in \mathbb{R}^m$ is the control input. The associated, general constraint can be formulated as

$$h(x) \geq 0, \tag{4.7}$$

for some continuously differentiable function, $h : \mathbb{R}^n \to \mathbb{R}$.

---

[2]Control-affine dynamics arise in many robotic systems, whose models are derived using Euler-Lagrange equations, as observed in [225]. As such, they can be used to describe a large variety of robotic platforms.

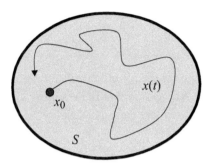

Figure 4.2: The set S is forward invariant with respect to the system dynamics in that, if the state starts in S, it remains in S for all future times.

In subsequent sections, we will discuss strategies and methods for making sure that such constraints are never violated, as well as anchor the constraint in Equation 4.7 in terms survivability for robots deployed in environments over long time-scales.

4.1.2 Remaining Safe Forever

So far, we have suppressed the explicit dependence of x on time, t. But, in order to make statements about what happens at certain points in time, we need to be a bit more precise in terms of expressing the desired constraint satisfaction properties. To this end, assume, initially, that we have a general, nonlinear system (with no control inputs),

$$\dot{x}(t) = F(x(t)), \tag{4.8}$$

where $x \in D$ for some open domain $D \subset \mathbb{R}^n$, and where F is locally Lipschitz continuous on D. Lipschitz continuity ensures that solutions exist uniquely, i.e., for any initial condition, $x_0 \in D$, such that $x(t_0) = x_0$, there exists a maximum time interval $I(x_0) = [t_0, T_{max})$, such that $x(t)$ is the unique solution to Equation 4.8 on the interval $I(x_0)$, e.g., [209]. What we are interested in is to understand if we can make sure that *if a system starts safe, it stays safe*. In other words, if it starts out in some safe set S, it remains in that set forever (or at least on the entire interval $I(x_0)$), as illustrated in Figure 4.2. This property is known as *forward invariance*, defined (using the notation from [11, 436]) as follows:

Definition 4.1 A set $S \subset \mathbb{R}^n$ is called *forward invariant* with respect to the dynamics in Equation 4.8, if, for every $x_0 \in S$, $x(t) \in S$, $\forall t \in I(x_0)$.

Gathering these observations together, what we need to do is solve the following problem.

Problem 4.1 Given the control-affine robot system in Equation 4.6, find a feedback law, $u(x)$, such that the resulting closed-loop system, $\dot{x} = f(x) + g(x)u(x)$, renders the super-level set

$$S = \{x \in \mathbb{R}^n \mid h(x) \geq 0\} \tag{4.9}$$

forward invariant, given the continuously differentiable function $h : \mathbb{R}^n \to \mathbb{R}$.

Throughout this chapter, we will refer to S as the *safe set*, and if $h(x) \geq 0$ represents that the robot is staying away from situations where it could never recover, i.e., it completely encodes the survivabiliy of the robot,[3] then ensuring forward invariance of the super-level set in Equation 4.9 captures exactly the ethos of long-duration autonomy. However, ensuring forward invariance is not an entirely straightforward matter, as will be seen in subsequent sections.

4.1.3 Nagumo and the Comparison Lemma

Since we assumed that the function $h : \mathbb{R}^n \to \mathbb{R}$ is C^1, i.e., that it is continuously differentiable, we can ask what happens on the boundary of the super-level set in Equation 4.9, i.e., on the set $\partial S = \{x \in \mathbb{R}^n \mid h(x) = 0\}$. If the system was somehow prevented from passing through this boundary, we could, by continuity of solutions, draw the conclusion that it never leaves the safe set, S.

We first note that ∇h is orthogonal to the tangents to ∂S, and that the gradient always points in the direction in which h increases the most, e.g., [58], i.e., it points into the interior of the open set $\{x \in \mathbb{R}^n \mid h(x) > 0\}$. In other words, if the inner product between ∇h and F in Equation 4.8 satisfies

$$\langle F(x), \nabla h(x) \rangle \geq 0, \ \forall x \in \partial S \tag{4.10}$$

---

[3] To *completely* encode everything associated with the survivability of a robot in an unknown and dynamic environment through a single, scalar function is certainly a tall order. But, for now, we will proceed as if this was indeed doable. Chapter 7 will largely be devoted to the search for such a survival constraint.

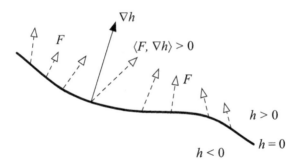

Figure 4.3: Since the inner product between ∇h and F is non-negative, the system will, by Nagumo's Theorem (Theorem 4.1), never leave the safe set where $h \geq 0$.

then, on the boundary, $F(x)$ also points into the interior of the safe set,[4] and the system will never flow in a direction that renders h negative. This holds as long as h is *regular*, i.e., when $\nabla h \neq 0$ on ∂S, as shown in Figure 4.3, and established in different forms in [1, 52, 55, 60, 290].

Put more carefully and precisely, and phrased in terms of [1] rather than Nagumo's original formulation in [290] to better fit the notation in this book, the corresponding theorem states:

Theorem 4.1 — Nagumo's Theorem. Given a dynamical system, $\dot{x} = F(x)$, with $x \in D \subset \mathbb{R}^n$ and F locally Lipschitz continuous on D. Let the set S be the super-level set of a continuously differentiable, regular function, $h : \mathbb{R}^n \to \mathbb{R}$, i.e., $S = \{x \in \mathbb{R}^n \mid h(x) \geq 0\}$, with $\nabla h(x) \neq 0$ whenever $h(x) = 0$, and where $S \subseteq D$. The set S is forward invariant if and only if $\langle \nabla h(x), F(x) \rangle \geq 0, \ \forall x \in \partial S$.

One attempt at ensuring forward invariance of the safe set in Equation 4.9 is thus to pick the closed-loop control, $u(x)$, such that the following condition holds (assuming that $h(x_0) \geq 0$, given the initial condition x_0),

$$
\begin{aligned}
&u \text{ is free to be whatever, when } h(x) > 0 \\
&u \text{ satisfies } \langle f(x) + g(x)u, \nabla h(x) \rangle \geq 0, \text{ when } h(x) = 0.
\end{aligned} \tag{4.11}
$$

---

[4]Provided the inequality is strict in Equation 4.10; otherwise it is tangential to ∂S.

This is, however, not a particularly inspired choice as the sudden nature of the constraint means that u will go from being whatever to all of a sudden being forced to satisfy a particular inequality constraint. And, there certainly are no guarantees that there are any solutions to this constraint. A better choice might be to approach the boundary set ∂S in a more careful and deliberate manner as a way of overcoming these issues. To this end, we need to understand a bit better what happens when $h > 0$, and not only focus on the case when $h = 0$. However, Nagumo's Theorem will still inform us about what needs to transpire on ∂S.

Armed with the knowledge that we need $\langle f(x) + g(x)u, \nabla h(x) \rangle \geq 0$ on the boundary when $h(x) = 0$, what should happen in the interior, i.e., when $h(x) > 0$? Assume for a moment that $x(t_0) = x_0$, with $h(x_0) > 0$, and that we had another function, $\hat{h}(t)$, with initial condition $\hat{h}(t_0) = h(x_0) > 0$, that asymptotically converged to the origin, i.e.,

$$\lim_{t \to \infty} \hat{h}(t) = 0.$$

One way of achieving this asymptotic convergence to the origin is to have $\hat{h}(t)$ satisfy the following differential equation

$$\dot{\hat{h}} = -\alpha(\hat{h}),$$

where $\alpha : \mathbb{R} \to \mathbb{R}$ is a locally Lipschitz continuous, extended class $\mathcal{K}_\infty$-function, i.e., α is a strictly increasing function through the origin ($\alpha(0) = 0$).[5] With this choice of dynamics for $\hat{h}$, we not only have $\hat{h} \to 0$ but also $\hat{h}(t + T) < \hat{h}(t)$, for all $T > 0$, i.e., it monotonically decreases down to zero from above, e.g., [14, 29, 209, 436].

Given such a function, $\hat{h}$, that decays asymptotically down to zero from above, what can one do with this function? If we somehow could ensure that h was greater than (or equal to) $\hat{h}$, then we would be able to establish that $h \geq 0$. To arrive at this result, we can make use of the Comparison Lemma (formulated in terms of $\geq$ rather than the typical $\leq$), that can be found, for example, in [366]:

---

[5]Technically speaking, it is enough if α belongs to class $\mathcal{K}$ rather than class $\mathcal{K}_\infty$, as shown in [14]. In the former case, one has to pay attention to the domain over which α is defined, while in the latter case, the function is defined over the entire real line. This makes things a bit easier from a bookkeeping point of view, which is why we go with this option.

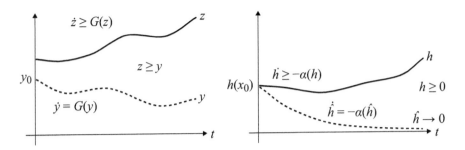

Figure 4.4: The Comparison Lemma ensures that $h(x(t)) \geq 0$, $\forall t \geq 0$, i.e., that if the system starts safe (in the set S), it remains safe for all times.

Lemma 4.1 — Comparison Lemma. Let $G : \mathbb{R} \to \mathbb{R}$ be locally Lipschitz continuous everywhere, and assume that solutions exist $\forall t \geq t_0$ to the differential equation,

$$\dot{y} = G(y), \quad y(t_0) = y_0.$$

Moreover, let $z(t)$ be a continuously differentiable function satisfying

$$\dot{z} \geq G(z), \quad z(t_0) \geq y_0.$$

Then $z(t) \geq y(t)$, $\forall t \geq t_0$.

The Comparison Lemma thus provides the machinery needed to ensure that $h \geq \hat{h}$, since we have that $h(x(t_0)) = \hat{h}(t_0)$ and $\dot{h}(x(t)) \geq -\alpha(h(x(t)))$. But, as $\hat{h} \to 0$ from above, we have the even stronger statement that $h \geq \hat{h} \geq 0$. In other words, if, in addition to being an extended class $\mathcal{K}_\infty$-function, α is locally Lipschitz continuous, $h(x(t)) \geq 0$, $\forall t \geq t_0$, can be ensured if $h(t_0) \geq 0$ and

$$\dot{h}(x) = \frac{\partial h(x)^T}{\partial x} \dot{x} = \frac{\partial h(x)^T}{\partial x} F(x) \geq -\alpha(h(x)) \tag{4.12}$$

along trajectories of the system, e.g., [11]. This is illustrated in Figure 4.4, and it is in fact exactly what is needed to ensure that S is forward invariant. Additionally, it will allow us to focus on survival *constraints* when designing robot controllers for long-duration deployments.

4.2 Control Barrier Functions

One of the takeaways from the previous section is that we need to manipulate $\dot{h}$ in order to ensure forward invariance. And, given the control-affine
robot model in Equation 4.6, along trajectories of the system, $\dot{h}$ takes on the
following form

$$\dot{h}(x) = \frac{\partial h(x)^T}{\partial x}\dot{x} = \frac{\partial h(x)^T}{\partial x}(f(x) + g(x)u).$$

Or, using Lie derivative notation [366], where $L_f h$ and $L_g h$ denote the
derivatives of h along f and g, respectively,

$$L_f h(x) = \frac{\partial h(x)^T}{\partial x}f(x), \quad L_g h(x) = \frac{\partial h(x)^T}{\partial x}g(x), \tag{4.13}$$

the time derivative of h becomes

$$\dot{h}(x) = L_f h(x) + L_g h(x)u. \tag{4.14}$$

Combining the expression in Equation 4.14 with the constraint in Equation 4.12 takes us to the following definition:

> **Definition 4.2 — Control Barrier Function (CBF).** Given the control-
> affine system in Equation 4.6, where f and g are locally Lipschitz
> continuous everywhere, and a smooth function $h: \mathbb{R}^n \to \mathbb{R}$, with the cor-
> responding super-level set $S \subseteq \mathbb{R}^n$. The function h is a *Control Barrier
> Function* (CBF) if there exists a Lipschitz continuous, extended class
> $\mathcal{K}_\infty$-function, α, such that
>
> $$\sup_u\{L_f h(x) + L_g h(x)u + \alpha(h(x))\} \geq 0, \ \forall x \in \mathbb{R}^n. \tag{4.15}$$

Given a CBF, h, the *admissible control space*, $K(x)$, can now be
defined as

$$K(x) = \{u \in \mathbb{R}^m \mid L_f h(x) + L_g h(x)u + \alpha(h(x)) \geq 0\}, \ x \in \mathbb{R}^n. \tag{4.16}$$

Note, first, that in these definitions, $\mathbb{R}^n$ can be replaced by an open domain
$D \subset \mathbb{R}^n$, depending on what set the dynamics are defined over. Second, the
formulation in Definition 4.2 is sometimes referred to as a Zeroing Control

Barrier Function (ZCBF) rather than a CBF to explicitly call out the fact that it allows for x to reach the boundary where $h(x) = 0$, as shown in [14, 436].

This is all we need to ensure that the super-level set to h, S, is forward invariant through the following key theorem, borrowed from [11, 14]:

Theorem 4.2 — Forward Invariance. Let $S \subset \mathbb{R}^n$ be the super-level set of a continuously differentiable, regular function $h : \mathbb{R}^n \to \mathbb{R}$. If h is a control barrier function, then any locally Lipschitz continuous controller in the admissible control space, $u(x) \in K(x)$, for the control-affine system in Equation 4.6, where f and g are locally Lipschitz continuous, renders the set S forward invariant. Additionally, the set S is asymptotically stable under this choice of control input in the sense that, asymptotically, x will reach S even if it, initially, starts outside of S.

This result is exactly what is needed to turn the vague statement of $x \in \mathcal{G}_{survive}$ into something with crisp mathematical meaning, and that is formulated in terms of a constraint in u—the control input—as opposed to a constraint in the state, x, which makes it an actionable constraint. To this end, we will use the shorthand notation

$$S_h[x; u] = L_f h(x) + L_g h(x) u + \alpha(h(x)) \tag{4.17}$$

to denote the safety constraint

$$S_h[x; u] \geq 0. \tag{4.18}$$

One should note that although this may look like a complicated expression, all the complicated terms are functions of x, i.e., of the state of the system which we, for now, assume we can access directly. The one variable that really matters is the decision variable, u, and the constraint is *linear* in u, which bodes well for future computational attempts at actually finding such a u.

4.2.1 Optimization-Based Control

Now that we have a way of encoding safety constraints using control barrier functions, we need some method of combining the constraints with the primary performance objectives. And, the solution to this problem has already been outlined through the (vague) constrained optimization formulation "min $\mathcal{C}_{task}$, subject to $x \in \mathcal{G}_{survive}$." We could thus formulate a general, instantaneous cost as a function of both state and control input, $\mathscr{C}_{task}(x, u)$, and solve the following problem:

$$\min_{u} \mathscr{C}_{task}(x, u)$$
$$\text{subject to } S_h[x; u] \geq 0, \tag{4.19}$$

where, as before, $h(x) \geq 0$ encodes the safety constraint.[6]

Although appealing in its simplicity, the formulation in Equation 4.19 suffers from the problem of only taking an instantaneous performance cost into account. This goes against virtually all of optimal control, where the state is projected into the future and the cost must be evaluated along the resulting state-input trajectories, e.g., [66, 230, 238]. The reason for this is that otherwise the controller can become myopic and way too focused on the near-term, at the expense of future performance. In fact, one can even construct examples where the controller may be optimal over a short time horizon, for any instant in time, yet the resulting system goes unstable due to the lack of a longer-term prediction [8].

A way of overcoming this hurdle and producing an instantaneous cost in order to make use of the CBF safety constraint formalism, yet require a longer term, predictive outlook on how the system should evolve, is to *not* use the performance cost, $\mathscr{C}_{task}$, in the constrained optimization problem. Instead, assume that some appropriate performance cost has already been defined using more traditional approaches. Ignoring the safety constraint, denote the (potentially unsafe) controller obtained from solving this problem by u_{nom}, where the subscript *nom* stands for "nominal." This controller, u_{nom}, could be obtained by solving the unconstrained optimal control problem repeatedly, at each time instant, making it a function of the state x, i.e., $u_{nom}(x)$, as depicted in Figure 4.5. Or it could be a one-shot solution, turning it into a function of time, $u_{nom}(t)$. Or it could not be optimal at all, but simply given by some other means, e.g., by an external user, as was done in [337, 338], or by some other regulator, as will be the case in Section 4.3.

Regardless of where u_{nom} comes from, what is done under the banner of safety critical control [14] is to modify this controller as little as possible in a least-squares sense, yet ensure that the modified controller satisfies the safety

---

[6]A similar formulation was used in [310] as a way of capturing other types of desired behaviors, including stability and tradeoffs between tasks with different priorities.

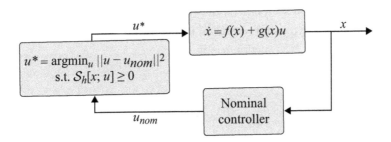

Figure 4.5: Optimization-based, safety critical control: The nominal controller is modified in a minimally invasive manner to ensure forward invariance of the safe set.

constraint. The resulting, actual controller, $u^\star$, is *minimally invasive* in the sense that it stays as close as possible to the nominal controller, subject to the safety constraint, i.e.,

$$u^\star = \mathrm{argmin}_{u \in \mathbb{R}^m} \|u - u_{nom}\|^2 \qquad\qquad (4.20)$$
$$\text{subject to } S_h[x; u] \geq 0.$$

One important aspect of the formulation in Equation 4.20 is that the cost is *quadratic* in the decision variable u, while the constraint is *linear* in u. In other words, this is a quadratic programming (QP) problem. In the absence of additional constraints on the control variable, such as saturation limits, the QP can be solved analytically. The resulting closed-form solution is given by the standard Karush–Kuhn–Tucker (KKT) conditions, e.g., [58]. This construction was utilized in [12, 152, 310] for the purpose of robust, safety critical control. And, even with additional constraints on u, the QP can still be solved effectively and is computationally unproblematic [58]. This is critical in real-time applications, such as robotics.

It should, in closing, be noted that the choice of extended class $\mathcal{K}_\infty$-function, α, matters in that this function regulates how quickly the state of the system can approach the boundary of the safe set S. Different choices of α lead to different behaviors near the boundary. For example, the choice of

$$\alpha(h) = \gamma h^{2p+1}, \quad p \in \mathbb{N}, \qquad\qquad (4.21)$$

satisfies the extended class $\mathcal{K}_\infty$-property for any non-negative integer p and positive real number $\gamma > 0$. The choice made in [11] is to simply let $\alpha(h) = h$.

However, to allow for a slightly more aggressive approach towards the boundary of the safe set, we will, for most of this book, go with the cube, i.e., to let the extended class $\mathcal{K}_\infty$-function be given by

$$\alpha(h) = \gamma h^3, \tag{4.22}$$

for some $\gamma > 0$.

4.2.2 Further Considerations

Control barrier functions and optimization-based control combine to provide a natural and computationally lightweight framework for encoding survival among robots deployed over long time-scales, tightly coupled to their "habitats." But, whenever something seems to good to be true, it probably is. In this case, the two distractions from the otherwise compelling CBF story are related to the *existence of solutions* and the potential for *deadlock*.

First, note that the main CBF result in Theorem 4.2 states that the safe set is rendered forward invariant *if* u satisfies the CBF constraint. But that is a big "if." What if there are no solutions to the constrained optimization problem in Equation 4.20? In that case, no guarantees can be given, and it is important to always add a feasibility-check to the to-do list in order to ensure that solutions do indeed exist. This is something that has to be done in addition to the formulation of the CBF, i.e., it is in general not something that is provided for free by the developed theory.

Second, even if u_{nom} in Equation 4.20 encodes something rather important, like progression towards a goal location, there are in general no guarantees that one always, by finding the minimizer to the QP, will progress towards that goal. Or even that one will progress at all. All that is ensured is safety. In fact, in [425] it was observed that when employing CBFs for the purpose of collision-avoidance, the robots would sometimes find themselves in situations where the only feasible solution to the QP was to do nothing. In other words, they ended up in a deadlock (albeit safe) situation. In [78], a solution to this problem was introduced that included a slight bias as a way of ensuring that the robots would always be able to negotiate collisions while bounding their velocities away from zero.

For now, neither of these two considerations should deter us from the use of CBFs. Instead, they should simply serve as a reminder that some additional care must be taken when designing safety controllers. This will be highlighted further in the next sections, where collision-avoidance controllers for multi-robot teams are investigated in some detail. But first, a collection of particularly relevant and pertinent survival constraints, hinted at in the previous chapter, will be covered as a bridge between long-term autonomy, survivability, and CBFs.

4.2.3 Survivability Constraints

One can envision a number of potential constraints that help promote robot longevity—some of which connect directly to the multi-robot behaviors in Chapter 2 and the ecological constraints in Chapter 3. We here discuss a few of these constraints, drawn primarily from [128]. This exposé will moreover provide an outlook towards topics to be discussed further in subsequent chapters of the book, culminating with the full-fledged survivability constraint in Chapter 7.

To make matters concrete, assume for simplicity that the state, x_i, of each robot is given in part by a directly controlled planar position, $p_i \in \mathbb{R}^2$. Additionally, as the availability of energy plays a crucial role to longevity, we add the energy level, E_i, to the state of each robot, i.e.,

$$x_i = \begin{bmatrix} p_i \\ E_i \end{bmatrix} \in \mathbb{R}^3, \tag{4.23}$$

where we, for now, go with a simplified dynamics (pure energy-decay),[7]

$$\dot{x}_i = \begin{bmatrix} u_i \\ \pm \kappa E_i \end{bmatrix}, \tag{4.24}$$

where $\kappa > 0$, and the sign in the energy dynamics depends on whether the robot is charging the batteries ($+$) or not ($-$). We moreover follow [13, 424] in choosing the extended class $\mathcal{K}_\infty$-function, α, according to Equation 4.22, as the cube $\alpha(h) = \gamma h^3$, for some $\gamma > 0$.

For this choice of system dynamics, we can now introduce a number of constraints that are fundamental to long-duration autonomy:

(i) *Separation* (based on the developments in [424] as well as on the separation term in Reynolds' Boids model):

---

[7] This energy model will have to become significantly more sophisticated and realistic in Chapter 5, where the explicit focus is on the "persistification" of robotic tasks.

Given a safety distance, $D_s > 0$, two robots, i, j, with positions p_i, p_j, are safe and sufficiently separated from each other as long as

$$h_{separation}(x_i, x_j) = \|p_i - p_j\|^2 - D_s^2 \geq 0. \qquad (4.25)$$

Under the model assumptions above, the corresponding CBF constraint, $S_{h_{separation}}[(x_i, x_j); (u_i, u_j)] \geq 0$, becomes

$$S_{h_{separation}}[(x_i, x_j); (u_i, u_j)] = (p_i - p_j)^T (u_i - u_j)$$
$$+ \frac{\gamma}{2}(\|p_i - p_j\|^2 - D_s^2)^3 \geq 0. \qquad (4.26)$$

(ii) Connectivity (as discussed in [423], and aligning with Reynolds' cohesion notion):

Two robots, i, j, are connected if they are close enough, i.e., if

$$h_{connect}(x_i, x_j) = D_c^2 - \|p_i - p_j\|^2 \geq 0, \qquad (4.27)$$

given a connectivity distance, $D_c > 0$. Or, formulated as a CBF constraint, $S_{h_{connect}}[(x_i, x_j); (u_i, u_j)] \geq 0$, where

$$S_{h_{connect}}[(x_i, x_j); (u_i, u_j)] = -(p_i - p_j)^T (u_i - u_j)$$
$$+ \frac{\gamma}{2}(D_c^2 - \|p_i - p_j\|^2)^3.$$

(iii) Coverage (based on the "locational cost" in [95], and described further in Chapter 2):

A robot that is tasked with monitoring a domain, $\mathscr{D}_i \subset \mathbb{R}^2$, where the importance of a point, $q \in \mathscr{D}_i$, is given by the density function $\phi(q) \in \mathbb{R}_+$, and where the quality of the sensor coverage associated with this point, as achieved by the robot located at $p_i \in \mathscr{D}_i$, can be encoded by the quality function $Q(\|p_i - q\|)$. (Typically, the farther away the point is, the worse the coverage is.) Additionally, let c be a lower limit for acceptable coverage performance, which gives the coverage constraint

$$h_{coverage}(x_i) = \int_{\mathscr{D}_i} Q(\|p_i - q\|)\phi(q)dq - c \geq 0. \qquad (4.28)$$

Note that, despite this looking like a rather complicated expression, the corresponding CBF constraint is still linear in the decision variable, u_i, i.e., it is actually not a particularly hard constraint to handle from a computational vantage point.

(iv) Spatial Anchoring (following the ecological notion of spatial anchoring, as per Chapter 3):

If E_i is the energy level associated with Robot i and $dist(p_i)$ is the distance to the closest "charging station" (as interpreted broadly), one can let $\mathcal{E}(dist(p_i))$ be the energy required to travel that distance under suitable assumptions on the motion model [374]. The requirement that the robot never gets stranded away from a charging station thus becomes

$$h_{anchor}(x_i) = (E_i - E_{min}) - \mathcal{E}(dist(p_i)) \geq 0, \qquad (4.29)$$

given some smallest acceptable energy level, E_{min}.

(v) Energy (based on the task persistification in [304], and elucidated upon in Chapter 5):

To prevent over-charging when a robot is present at a charging station, one would be tempted to introduce the constraint $h_{energy}(x_i) = E_{max} - E_i \geq 0$, where E_{max} is the maximum energy level supported. However, viewing h_{energy} as an output of the system, the relative degree associated with this output is not well-defined in that the input u_i never shows up in any of the time derivatives of h_{energy}, which makes it a poor choice for a constraint. The problem is that charging only happens when the robot is close to the charging station, say at a distance $\Delta > 0$, which is not reflected in the constraint. If we, as before, let $dist(p_i)$ be the distance to the charging station, one can define an additional, smooth approximation of a step function, $step_\Delta(dist(p_i))$, that is zero when $dist(p_i) \geq \Delta$, and 1 when $dist(p_i) = 0$. The energy constraint thus becomes

$$h_{energy}(x_i) = E_{max} - step_\Delta(dist(p_i))E_i \geq 0. \qquad (4.30)$$

As already noted, when handling CBF constraints, there might not exist any feasible solutions. This could for example be the case if a robot "wakes up" with almost completely depleted batteries too far from a charging station, in which case the CBF constraint associated with $h_{anchor} \geq 0$ would not have a solution. Or, there might simply not be enough density in the

domain, $\mathscr{D}_i$, for any point $x_i \in \mathscr{D}_i$ to satisfy $h_{coverage} \geq 0$. These are ultimately issues pertaining to the choice of constraints, and they may need to
be tweaked in order to fit the particulars of the scenarios under consideration.
Regardless, these types of constraints will be used throughout the remainder of the book in order to ensure that robots do not get stranded away from
charging stations without sufficient energy, while, at the same time, they remain safe, connected, or exhibit other types of relevant behaviors, such as
covering a sufficiently large domain. To achieve this, we must be able to
combine multiple CBFs into a single, unified survival CBF. This, however,
is a matter for subsequent chapters. Instead, we conclude this chapter with a
more in-depth discussion of how to properly handle collision-avoidance constraints when there are a large number of robots present, as well as highlight
how safe and curiosity-driven learning can be managed as part of the CBF
paradigm.

4.3 Collision-Avoidance

To showcase the use of CBFs to ensure safety using optimization-based
control, let us return to the collision-avoidance question discussed in the beginning of the chapter. But, rather than considering "bead-robots" on a string,
let the robots move around in the plane. The reason collision-avoidance deserves special consideration is that, at a minimum, any reasonable deployment
of robots in a complex, dynamic, and unknown environment will require that
the robots do not run into objects in the environment, nor into each other, e.g.,
[25, 68, 273, 276, 355, 362]. In fact, multi-agent robotics is typically focused
on achieving certain primary objectives, such as assembling and maintaining
formations, covering areas of interest, exploring environments, or tracking
boundaries, e.g., [36, 68, 69, 95, 276], with collision-avoidance added in afterwards, thereby making the constrained optimization formalism particularly
well-suited.[8]

---

[8]The idea of using collision-avoidance controllers that are *minimally invasive* in the
sense that they only modify the nominal controller when collisions are truly imminent has
been considered in a few different contexts. In [408], pairs of aircraft were considered,
where the individual aircraft switch between normal operation modes and evasive maneuvers that are guaranteed to be safe no matter what actions the other aircraft are taking.
A similar idea was presented in [9, 417], where a so-called velocity obstacle method was
used to calculate the optimal, safe velocity. Algorithms for avoiding imminent collisions
with static obstacles were developed in [149, 312, 313] by decelerating the robots to zero
velocities by applying a maximum braking force.

4.3.1 Centralized Safety Barriers

From a bookkeeping perspective, the foundation for ensuring that collisions are avoided is that *all* potential, pairwise robot-to-robot collisions are accounted for. As such, a natural first bid is to let a central unit keep track of all robot pairs and let it dictate how the nominal controllers should be modified in order to avoid collisions. But, as robots that are far away from each other are in no immediate danger of colliding, this is a needlessly pedantic approach. As such, it will have to be relaxed to only consider a significantly smaller subset of pairs, consisting of nearby robots. Additionally, the central unit will eventually have to go as well in order to empower the robots to make decisions themselves in real-time. For now, however, let us start with this overly circumspect approach and get to the improvements later.

Concretely, consider a multi-robot system consisting of N planar, mobile robots with dynamics

$$\begin{bmatrix} \dot{p}_i \\ \dot{v}_i \end{bmatrix} = \begin{bmatrix} 0 & I_{2\times 2} \\ 0 & 0 \end{bmatrix} \begin{bmatrix} p_i \\ v_i \end{bmatrix} + \begin{bmatrix} 0 \\ I_{2\times 2} \end{bmatrix} u_i, \quad i \in [N], \tag{4.31}$$

where $[N]$ is the index set, $\{1, \ldots, N\}$, $x_i = [p_i^T, v_i^T]^T \in \mathbb{R}^4$, with $p_i \in \mathbb{R}^2$, $v_i \in \mathbb{R}^2$, and $u_i \in \mathbb{R}^2$ being the positions, velocities, and inputs (acceleration commands) of Robot i, respectively. As before, we let the acceleration be bounded by $\|u_i\| \leq a$.

Following the route established for the "bead-robots," to avoid imminent collisions between moving robots, their relative velocity needs to safely be reduced to zero. To this end, let the relative position and velocity between Robots i and j be given by $\Delta p_{ij} = p_i - p_j \in \mathbb{R}^2$ and $\Delta v_{ij} = v_i - v_j \in \mathbb{R}^2$. Additionally, let $D_{ij} = \|\Delta p_{ij}\|$ be the distance between the robots. The change in distance, as established in [424], is given by the normal component, δv_{ij}, of the relative velocity, i.e., by the magnitude of the projection of Δv_{ij} onto Δp_{ij}. As illustrated in Figure 4.6, the normal component of the relative velocity is given by

$$\delta v_{ij} = \dot{D}_{ij} = \frac{\langle \Delta p_{ij}, \Delta v_{ij} \rangle}{\|\Delta p_{ij}\|} = \frac{\Delta p_{ij}^T}{\|\Delta p_{ij}\|} \Delta v_{ij} \in \mathbb{R}. \tag{4.32}$$

The rate of change in relative distance, δv_{ij}, is indeed what might lead to a collision between Robots i and j, while the tangential component of Δv_{ij} produces only distance-preserving rotations. If $\delta v_{ij} > 0$, the robots are moving away from each other, while a negative δv_{ij} means that they are getting closer.

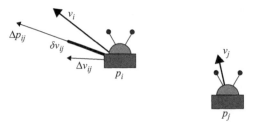

Figure 4.6: Relative position and velocity between two robots

The next order of businesses is to regulate δv_{ij} so that imminent collisions are avoided when the maximum relative braking force is applied. And, in a manner completely analogous to the one-dimensional case, based on the observation that the worst case scenario is when δv_{ij} is equal to $\|\Delta v_{ij}\|$ (the robots are driving straight towards each other), the two-dimensional case reduces to the one-dimensional case, as shown in [424]. Combining this observation with the one-dimensional constraint in Equation 4.5 gives that "slamming the brakes" will prevent a collision when

$$-\frac{\Delta p_{ij}^T}{\|\Delta p_{ij}\|}\Delta v_{ij} \leq 2\sqrt{a(\|\Delta p_{ij}\| - D_s)}, \ \forall \, i \neq j. \tag{4.33}$$

The benefit of this formulation is that it is constructive in the sense that "slamming the brakes" is always a feasible (and safe) solution. As such, existence of solutions is already baked in, and we have established the following, key property.

The pairwise, collision-free safe set, S_{ij}, is given by the super-level set

$$S_{ij} = \{(x_i, x_j) \in \mathbb{R}^8 \mid h_{ij}(x_i, x_j) \geq 0\}, \ \forall \, i \neq j, \tag{4.34}$$

to the CBF

$$h_{ij}(x_i, x_j) = 2\sqrt{a(\|\Delta p_{ij}\| - D_s)} + \frac{\Delta p_{ij}^T}{\|\Delta p_{ij}\|}\Delta v_{ij}. \tag{4.35}$$

Using the notational convenience of letting h_{ij} be shorthand for $h_{ij}(x_i, x_j)$, and setting $\Delta u_{ij} = u_i - u_j$, the (scaled) time derivative of h_{ij} is given by

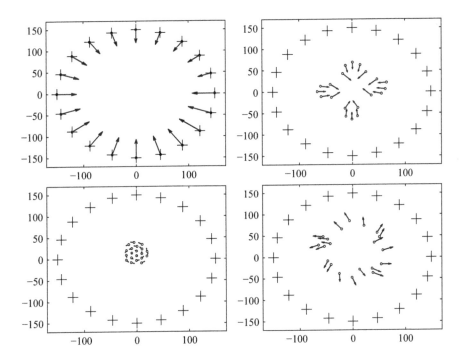

Figure 4.7: Simulation of a multi-robot position-swapping task, rendered safe by a centralized CBF. The circles and arrows represent the current positions and velocities of the robots, respectively. Source: Fig. 3 in Wang et al. (2017).

$$\|\Delta p_{ij}\|\dot{h}_{ij} = \Delta p_{ij}^T \Delta u_{ij} - \frac{(\Delta v_{ij}^T \Delta p_{ij})^2}{\|\Delta p_{ij}\|^2} + \|\Delta v_{ij}\|^2 + \frac{a\Delta v_{ij}^T \Delta p_{ij}}{\sqrt{a(\|\Delta p_{ij}\| - D_s)}}.$$
(4.36)

As such, the resulting CBF constraint, $\|\Delta p_{ij}\|\dot{h}_{ij} \geq -\|\Delta p_{ij}\|\gamma h_{ij}^3$, can be written as

$$S_{h_{ij}}[(x_i, x_j); (u_i, u_j)] = \Delta p_{ij}^T \Delta u_{ij} - \frac{(\Delta v_{ij}^T \Delta p_{ij})^2}{\|\Delta p_{ij}\|^2} + \|\Delta v_{ij}\|^2$$
$$+ \frac{a\Delta v_{ij}^T \Delta p_{ij}}{\sqrt{a(\|\Delta p_{ij}\| - D_s)}} + \gamma h_{ij}^3 \|\Delta p_{ij}\| \geq 0.$$
(4.37)

A simulation of the centralized safety barrier functions in action, applied to a team of 20 mobile robots, is shown in Figure 4.7. There, all robots start out equally spaced on a circle. The goal of the nominal controller is to make the robots swap positions with robots on the opposite side of the circle, which is a recipe for disaster as the robots will all have to go through the same central point in the middle of the circle. The nominal controller is a feedback control law,[9] and left to its own devices, the nominal controller will thus cause the robots to collide at the center of the circle.

The result of running the nominal, position-swapping controller, together with the CBF safety constraint in the optimization-based control setting in Equation 4.20, is displayed in Figure 4.7. As expected, when collisions are not imminent, the robots move towards the center, following the nominal controllers closely. As robots get too close to each other, the CBF constraints kick in and navigate the robots safely through the crowded, central region.

4.3.2 Decentralized Safety Barriers

The centralized safety barrier function in Equation 4.37 considers *all* pairs of robots as an overly conservative precaution. Additionally, if an optimization-based safety controller were to be used, as in Figure 4.5, it would have to be managed by an all-knowing computational entity. Topologically speaking, what is required is an all-to-all interaction graph, which does not scale gracefully in the number of robots. This clearly can be improved upon as collisions can only materialize between nearby robots. What is needed is an understanding of just how close "nearby" needs to be for it to be worthwhile to consider the pairwise robot-to-robot interaction.

The way to establish a bound for how close robots need to be for their pairwise interactions to be worth keeping track of is to find a distance $\bar{\Delta}$, beyond which they are always safe no matter what the control action is, i.e., a distance beyond which $\dot{h}_{ij} \geq -\gamma h_{ij}^3$ for all choices of input. To this end, recall that $D_{ij} = \|\Delta p_{ij}\|$ and reformulate h_{ij} in Equation 4.35 in terms of D_{ij} and $\dot{D}_{ij}$ as

$$h_{ij} = \dot{D}_{ij} + 2\sqrt{a(D_{ij} - D_s)}. \tag{4.38}$$

The derivative of h_{ij} is furthermore given by

$$\dot{h}_{ij} = \ddot{D}_{ij} + \sqrt{\frac{a}{D_{ij} - D_s}}\dot{D}_{ij}. \tag{4.39}$$

---

[9]The particular, nominal controller used in this scenario is given by the PD (Proportional-Derivative) controller $u_{nom,i} = -k_1(p_i - r_i) - k_2 v_i$, which drives the robots to their target positions, r_i, $i \in [N]$.

Now, if the robots could accumulate arbitrarily large speeds, no amount of separation would be safe. This is clearly a physical impossibility. As such, assume that the velocities are bounded as well, i.e., that $\|v_i\| \leq b$. With these velocity and (already imposed) acceleration limits, lower bounds on h_{ij} and $\dot{h}_{ij}$ can be derived, as was done in [424], by considering the worst case scenario ($\ddot{D}_{ij} = -2a$, $\dot{D}_{ij} = -2b$), yielding

$$h_{ij} \geq 2\sqrt{a(D_{ij} - D_s)} - 2b \tag{4.40}$$

$$\dot{h}_{ij} \geq -2a - 2\sqrt{\frac{a}{D_{ij} - D_s}} b. \tag{4.41}$$

Now, if we impose the distance requirement that pairwise robots are only considered if $D_{ij} \geq \bar{\Delta}$, where

$$\bar{\Delta} = D_s + \frac{1}{4a} \left(\sqrt[3]{\frac{4a}{\gamma}} + 2b \right)^2, \tag{4.42}$$

we get that, as long as $D_{ij} > \bar{\Delta}$, the following inequalities hold

$$\sqrt{4a(D_{ij} - D_s)} > 2b, \quad \text{and} \quad h_{ij} > \sqrt[3]{\frac{4a}{\gamma}}.$$

Therefore,

$$\dot{h}_{ij} \geq -4a \geq -\gamma h_{ij}^3. \tag{4.43}$$

What this means is that no matter what control action Robot j takes, it always satisfies the pairwise safety barrier constraint in Equation 4.37 with Robot i. Therefore, there is no need for Robot i to consider Robot j. We summarize this below.

> In order to avoid collisions, Robot i only needs to consider a neighboring robot (Robot j) in the collision-avoidance CBF constraint in Equation 4.35 if $D_{ij} \leq \bar{\Delta}$, where $\bar{\Delta}$ is given by Equation 4.42.

The one matter that needs to be resolved for the collision-avoidance story to come to its final resolution is to establish what the robots themselves should be doing. So far, the control decisions are taken by a central, computational unit. But that will have to change. Returning to the pairwise CBF constraint in Equation 4.37, we note that it can be written as

$$\Delta p_{ij}^T(u_i - u_j) \geq \beta_{ij}(x_i, x_j), \tag{4.44}$$

for some function $\beta_{ij}(x_i, x_j)$. As such, we could simply distribute the responsibility of maintaining this pairwise constraint equally across the two robots in the sense that Robot i needs to ensure that

$$\Delta p_{ij}^T u_i \geq \frac{1}{2}\beta_{ij}(x_i, x_j) \tag{4.45}$$

while Robot j ensures that

$$\Delta p_{ij}^T u_j \leq -\frac{1}{2}\beta_{ij}(x_i, x_j). \tag{4.46}$$

We note that an equivalent way of formulating the latter of these two constraints is

$$\Delta p_{ji}^T u_j \geq \frac{1}{2}\beta_{ji}(x_j, x_i), \tag{4.47}$$

i.e., the constraint is symmetric and the robots themselves do not need to keep track of whether they are Robot i or Robot j in the above expressions.

As a result of this division of labor, it is possible to relegate all control decisions to the individual robots themselves. What they should do is only consider robots that are closer to them than $\bar{\Delta}$. And for those robots, they should add the CBF constraint in Equation 4.45 to their own, individual constrained optimization problems. To observe this strategy in action, Figure 4.8 shows the results from an experiment where real robots execute the decentralized, constraint-based control strategy that allows them to swap positions with each other in an orderly and collision-free manner.

4.4 Safe Learning

As another example of the utility of a constraint-based approach to robotics, consider the problem of learning to perform new tasks in a safe manner. Just as animals "invest" in the future by caring for offspring at the expense of more immediate rewards, e.g., [102, 107], robots on long-duration deployments may find themselves in between missions. The first order of business

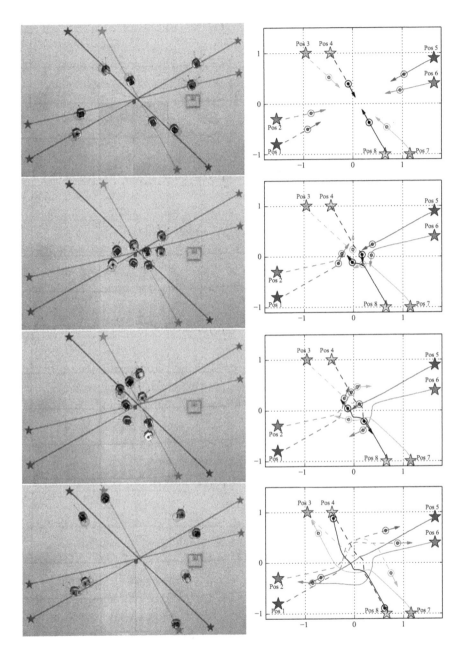

Figure 4.8: Experiment with eight mobile robots swapping positions. The pictures on the left are taken with an overhead camera. The stars and lines representing the target positions and pairs of swapped positions are projected onto the floor using a projector. The figures on the right display the positions, velocities and trajectories of the robots. Source: Fig. 3 in Wang et al. (2017).

for the robots should always be their survival. But, as observed already in Chapter 2, why not take advantage of the free time to learn new things?

Central to machine learning in general, and reinforcement learning in particular, is the idea that by leveraging prior data, current and future performances can be improved, e.g., [202, 288, 398]. To obtain rich and meaningful data to learn from, the world (external as well as internal) must be sufficiently well explored. But how can the required exploration be combined with safety in a way that does not overly restrict the actions the robots may take? Or, phrased slightly differently, how can sufficient coverage of the state and control spaces be ensured in order to facilitate learning without jeopardizing the robots themselves? We here, briefly, show one example of how the control barrier function framework can be put to use also in this context.

To focus the discussion and render the learning task under consideration concrete, assume that what needs to be learned is some unknown aspect of the dynamics, i.e., there is an unknown part, $d(x)$, added to the control-affine form in Equation 4.6,

$$\dot{x} = f(x) + g(x)u + d(x). \tag{4.48}$$

This formulation could capture factors such as aerodynamic wind effects experienced by unmanned aerial vehicles, as in [347], friction forces or other ground effects [131], or unknown currents influencing the motions of underwater vehicles, e.g., [400].

To establish an approximation of $d(x)$, assume that we have made M noisy measurements of the time derivative of the system,

$$\omega_q = \dot{x}_q + \eta_q, \quad q \in [M], \tag{4.49}$$

where the subscript refers to the measurement index, with ω_1 being the first measurement, and so on, and where we assume that the measurement noise, η_q, belongs to $\mathcal{N}(0, v^2)$, i.e., that it is normally distributed with zero mean and variance v^2. By subtracting away the known parts of the dynamics, and assuming that the state of the system, x_q, $q \in [M]$, is fully known when the measurements are made—obviously, the control inputs are known as well—a sequence of estimates of the unknown dynamics is obtained,

$$\hat{d}(x_q) = \omega_q - f(x_q) - g(x_q)u_q, \quad q \in [M]. \tag{4.50}$$

Now, the question is how to turn these pointwise, noisy measurements into a full-blown function approximation, $\hat{d}(x)$. The result will depend on what structure one wants to impose on $d(x)$, and one convenient choice of model is that of a Gaussian Process (GP), e.g., [5, 45, 113], which lets every point in the state space be associated with a normally distributed random variable. In particular, if we assume (purely for the sake of notational simplicity[10]) that $n = 1$, i.e., the state is scalar, the GP belongs to the class $GP(0, k(x, x'))$, with a prior mean of zero, and the covariance depending on the (Gaussian) kernel function, $k(x, x')$, that encodes the degree of similarity between states x and x'.

Since any finite number of data points form a multivariate normal distribution [185, 350], one can obtain the posterior distribution of $d(x)$ by conditioning on the measurement sequence, as was done in [426]. In particular, given the measurement sequence in Equation 4.50, the posterior distribution of $d(x^*)$, at query state x^*, has mean and variance

$$m(x^*) = k^{*T} (K + v^2 I)^{-1} \hat{d},$$
$$\sigma^2(x^*) = k(x^*, x^*) - k^{*T} (K + v^2 I)^{-1} k^*, \tag{4.51}$$

where $\hat{d} = [\hat{d}(x_1), \hat{d}(x_2), \ldots, \hat{d}(x_M)]^T$, K is the $M \times M$ kernel matrix, with $K_{i,j} = k(x_i, x_j)$, and $k^* = [k(x_1, x^*), k(x_2, x^*), \ldots, k(x_M, x^*)]^T$.

As shown in [426], this construction can be used to provide a high-probability confidence interval for the unmodeled dynamics, $d(x)$, as

$$\mathscr{D}(x) = \{d \mid m(x) - k_\delta \sigma(x) \le d \le m(x) + k_\delta \sigma(x)\}, \tag{4.52}$$

where k_δ is a design parameter to get to $(1 - \delta)$ confidence, for some design choice, $\delta \in (0, 1)$. For instance, 95.5% and 99.7% confidences are achieved at $k_\delta = 2$ and $k_\delta = 3$, respectively [185, 350, 426].

The use of GPs for approximating unknown aspects of the dynamics is interesting in its own right. But, it does not quite fit the theme thus far. In particular, it is not clear what this has to do with constraints and barrier functions. Or with the idea of "safe learning." To illuminate these connections, we have to return to the exploration issue and ask ourselves how the measurements, $\hat{d}(x_q)$, $q \in [M]$, were generated in the first place. In other words, we need to consider the problem of how to pick the measurement locations, x_q, $q \in [M]$, in such a way that we learn as much as possible about the system, yet, at the

---

[10]If $n > 1$, multivariate versions can be readily deployed, e.g., [426]. Nothing technical really changes beyond the need for slightly more involved notational bookkeeping.

same time, ensure that it does not stray into unsafe regions from where the system cannot be prevented from going unstable.[11]

It should be noted that this search for the unmodeled dynamics, $d(x)$, coupled with an identification of safety with stability, are particular choices of learning objectives that explicitly call out the idea of safe learning. They can, however, be generalized to encompass other objectives, e.g., [5], and we make these choices primarily for the purpose of highlighting how barrier functions can be put to effective use also in learning applications. In fact, to focus the problem formulation further, what we will do is proceed to learn a barrier function, $h(x)$, that lets us maximize the size of the safe set. This set is, by design, rendered forward invariant, i.e., it captures a region of the state-space from which the system can never be forced away by unmodeled dynamical effects. This is, as we will see, particularly pertinent to aerial vehicles that must learn how to manage unknown airflows without being swept out of the air and crashing.

4.4.1 Learning Barrier Functions

In light of the previous discussion, what we want to do is find the barrier function, $h(x)$, that produces the largest possible safe set, i.e., the largest possible region that can be rendered forward invariant.[12] More concretely, the goal of the learning process is to maximize the volume of the *barrier-certified, safe region*, by which we mean the set $\mathscr{C}_h = \{x \mid h(x) \geq 0\}$ that has been rendered forward invariant.[13] This, in turn, is achieved by incrementally adjusting $h(x)$ as better understandings of the unknown dynamics is obtained through the measurements, i.e.,

---

[11]If one equates "unsafe" with "unstable," this question of ensuring safety in terms of stability throughout the learning process has been investigated more thoroughly, for example, in [5, 45, 210, 351, 426].

[12]The connection between safety and stability is made explicit by observing that the unsafe region is the region where it is no longer possible to remain, no matter what the control input is.

[13]Since maximizing the volume of $\mathscr{C}_h$ is a nonconvex problem in general, we can restrict $\mathscr{C}_h$ somewhat by parameterizing the barrier function, $h_\mu(x)$, with μ being the parameter, in order to simplify the problem. For example, in [426], $h_\mu(x)$ was given by $1 - Z(x)^T \mu Z(x)$, where $Z(x)$ is a vector of monomials, and μ is a positive semi-definite matrix. This parameterization makes the maximization of vol($\mathscr{C}_h$) equivalent to the minimization of the trace of μ.

$$\max_{h(\cdot)} \text{vol}(\mathscr{C}_h)$$

$$\text{subject to } \max_{u} \min_{d \in \mathscr{D}(x)} \left\{ \frac{\partial h}{\partial x}(f(x) + g(x)u + d) + \alpha(h(x)) \right\} \geq 0, \ \forall x \in \mathscr{C}_h. \tag{4.53}$$

Note that this constraint[14] is phrased in terms of $\mathscr{D}(x)$ rather than the unknown $d(x)$, which means that it is possible to actually evaluate the constraint. However, the price one has to pay for this convenience is a *probabilistic* rather than deterministic safety guarantee.

Since u and d are independent from each other, the max-min constraint can be rewritten as

$$\max_{u} \left\{ L_g h(x) u \right\} + \min_{d \in \mathscr{D}(x)} \left\{ \frac{\partial h}{\partial x} d \right\} + L_f h(x) + \alpha(h(x)) \geq 0, \ \forall x \in \mathscr{C}_h. \tag{4.54}$$

Using the high confidence interval, $\mathscr{D}(x)$, in Equation 4.52, the problem expands to

$$\max_{h(\cdot)} \text{vol}(\mathscr{C}_h)$$

$$\text{subject to } \max_{u} \left\{ L_g h(x) u \right\} + \frac{\partial h}{\partial x} m(x) - k_\delta \left| \frac{\partial h}{\partial x} \right| \sigma(x) + L_f h(x) \tag{4.55}$$

$$+ \alpha(h(x)) \geq 0, \ \forall x \in \mathscr{C}_h,$$

which takes us to the following result, based on the developments in [426]:

Lemma 4.2 If u is selected so as to satisfy

$$L_g h(x) u + \frac{\partial h}{\partial x} m(x) - k_\delta \left| \frac{\partial h}{\partial x} \right| \sigma(x) + L_f h(x)$$

$$+ \alpha(h(x)) \geq 0, \ \forall x \in \mathscr{C}_h, \tag{4.56}$$

---

[14]Due to the assumption that $x \in \mathbb{R}$, the derivative of h is a scalar as well, which is why we do not need to keep track of the transpose, as is the case in higher-dimensional situations.

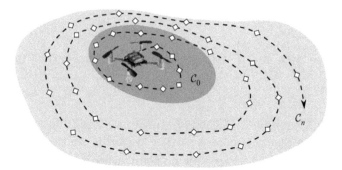

Figure 4.9: Incremental learning of barrier functions. The smaller region, $\mathscr{C}_0$, and the larger region, $\mathscr{C}_n$, are the barrier-certified, safe regions obtained initially and after n measurements of $\hat{d}$, respectively. The key feature in this pictorial representation of the process is the fact that the barrier-certified, safe region gradually expands as more and more data points are collected.

---

then the safety barrier constraint

$$\frac{\partial h}{\partial x}(f(x) + g(x)u + d(x)) + \alpha(h(x)) \geq 0 \tag{4.57}$$

is satisfied by this u, for all $x \in \mathscr{C}_h$, with probability $(1 - \delta)$, which is the confidence associated with the high-probability confidence interval, $\mathscr{D}(x)$, in Equation 4.52.

As more data points are accumulated, the uncertainty, $\sigma(x)$, gradually decreases. As a result, more states will satisfy the barrier constraint, as is pictorially illustrated in Figure 4.9. What this means is that the exploratory aspect of the learning process corresponds to a purposeful collection of data points in order to reduce $\sigma(x)$, while maximizing the volume of $\mathscr{C}_h$.

Since this process is not associated with any particular task beyond the reduction of uncertainty, it adheres to the curiosity-driven learning program, discussed in Chapter 2; what was informally described in Figure 2.12 has now been placed on sound mathematical footing. In particular, in order to increase the learning efficiency during the exploration phase, the most uncertain state in $\mathscr{C}_h$ could be sampled, i.e., one could select the next aspirational sample point as

$$x_{\text{next}} = \underset{x \in \mathscr{C}_h}{\arg\max} \, \sigma(x). \tag{4.58}$$

If we assume that a nominal exploration controller, $u_{explore}$, can be designed to drive the system from the current state, x, to x_{next} in Equation 4.58, then the optimization-based, safety procedure in Figure 4.5 can be invoked. In other words, the controller that ultimately determines the location of the next data point should be given by

$$\min_u \|u - u_{explore}\|^2$$

$$\text{subject to}\quad L_g h(x)u + + \frac{\partial h}{\partial x} m(x) - k_\delta \left| \frac{\partial h}{\partial x} \right| \sigma(x)$$

$$+ L_f h(x) + \alpha(h(x)) \geq 0, \tag{4.59}$$

which we can rewrite more compactly as

$$S_h[x; u] \geq k_\delta \left| \frac{\partial h}{\partial x} \right| \sigma(x) - \frac{\partial h}{\partial x} m(x). \tag{4.60}$$

The resulting exploration thus stays as close as possible to the nominal controller, $u_{explore}$, while always honoring the safety requirements. The way this gets implemented is that an inner loop executes the exploratory controller (the curiosity part) in Equation 4.59. Whenever x is close enough to x_{next}, or if the inner loop times out, a new measurement is taken, $\sigma(x)$ and $m(x)$ are updated, as per Equation 4.51, and a new $h(x)$ (and correspondingly, $\mathscr{C}_h$) is produced as a result of the volume maximization in Equation 4.55. The program terminates when the barrier-certified, safe region, $\mathscr{C}_h$, stops expanding through the iterative, volume maximization process.

4.4.2 Applications to Aerial Robotics

To illustrate its operations, the GP-based, curiosity-driven, safe learning algorithm is deployed on both a simulated and actual quadrotor. In simulation, the real weight of the quadrotor is 1.4 times the weight used by the model. In addition, an unknown, constant wind of $0.1g$ is affecting the quadrotor, as shown in Figure 4.10. Since the standard fixed-pitch quadrotor cannot generate reverse thrust, the thrust control is limited to the interval $[-1.8mg, 0]$. This setup is very challenging, as the learning-based quadrotor controller needs to deal with both a significantly inaccurate model as well as limited thrust capabilities.

To see how the learning process improves the performance, consider the problem of having the quadrotor track a reference trajectory, as shown in Figure 4.10. To make an informative comparison, the position errors in three

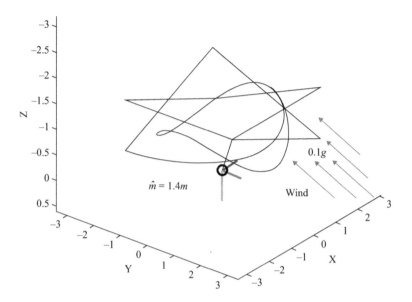

Figure 4.10: A simulated quadrotor flies in an unknown wind field using a highly inaccurate model.

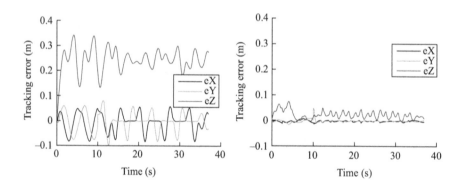

Figure 4.11: Tracking error when tracking a reference trajectory using a nominal model (Left) and after learning the unknown dynamics (Right). Source: Fig. 5 in Wang (2018).

dimensions are considered for the case when no learning takes place, as shown in Figure 4.11 (Left) and when the GP-based, safe learning algorithm in Equation 4.55 has terminated, as shown in Figure 4.11 (Right). As can be seen, the tracking error associated with the learning-based controller is significantly smaller than the tracking error obtained without learning.

Figure 4.12: By incrementally probing the airflow around a fan, the quadrotor learns how to fly through the blade-less fan in a safe manner. Additionally, a GP model is produced as part of this process, describing the effect the fan has on the quadrotor dynamics.

The same GP-based, curiosity-driven, safe learning algorithm used in simulation is deployed also on an actual, miniature quadrotor, tasked with flying through a blade-less fan. The undisturbed model corresponds to the case when the fan is off, i.e., the external airflow is virtually nonexistent. Once the fan turns on, the quadrotor must learn how the unknown, but significant, airflow influences its motion in order to successfully fly through the fan. The result from this experiment, based on the developments in [426], is shown in Figure 4.12.

5   Persistification of Robotic Tasks

Empowered by the control barrier formalism for ensuring constraint-satisfaction, we are now able to tackle the original long-duration constraint discussed already in Chapter 1, where a robot has to circle back to a charging location, every now and then, to recharge. What is needed is a purposeful restriction of the robot's nominal behavior in such a way that it always has sufficient energy on hand to be able to reach the designated charging area. If this recharging strategy was to coexist with a successful execution of a primary task, we would say that this task has been *persistified* in that it can be carried out indefinitely. The reason why this particular situation deserves its own, stand-alone chapter, as opposed to being treated as one constraint among others, is that persistification is a fundamental, necessary condition for robot longevity [304].

The primary tasks we will focus on in this chapter, as motivating scenarios, are environmental monitoring and exploration.[1] Not only do these tasks connect well with the environmental theme invoked by the robot ecology

---

[1]The persistification framework will have to be agnostic to the particulars of the primary tasks in order to support a wide range of long-duration deployment scenarios. As such, it will not rely on the environmental monitoring interpretation for its mathematical composition.

appellation, they are also prime examples of robotic tasks that have to be executed over long time horizons. However, as robots can only store and carry a limited amount of energy in their batteries, clever hardware solutions alone are not sufficient to allow the robots to perform these types of monitoring tasks perpetually, e.g., [287], which was one of the key attributes of a long-duration deployment identified in Chapter 1. To this end, we adhere to the story line from the previous chapter by minimally modifying the nominal control inputs that define the monitoring tasks[2] in order to enter the long-duration autonomy realm. As a result, the robots are allowed to freely execute the monitoring tasks whenever they have enough energy stored in their batteries; whereas they are forced to recharge whenever their energy caches are running low.

5.1 Energy Dynamics

In order to frame the opportunities afforded by the environment in terms of recharging possibilities, one needs to not only model the robot, but also the coupling between robot and environment. In other words, in addition to needing a representation of the robot dynamics and of how batteries deplete over time, what is needed is a description of the robot-environment interactions in terms of the recharging mechanism supported by different environmental conditions and configuration.

5.1.1 Environmental Interactions

Let the environment, i.e., the domain in which the robots are deployed, be given by the compact set $\mathscr{E} \subset \mathbb{R}^d$, where d is the dimension in which the robots operate, i.e., $d = 2$ in the case of ground robots, and $d = 3$ for aerial vehicles. As the full state of the robot is typically not given in terms of its position in $\mathbb{R}^d$, we let the output function,

$$\pi : \mathbb{R}^n \to \mathscr{E}, \tag{5.1}$$

---

[2]Environmental monitoring [143, 166, 233, 319, 384], exploration [69, 219, 309], and coverage [95, 218, 229] are well-studied topics in robotics. Persistent variants have also been considered, e.g., in [281], which examines persistent coverage through the lens of a vehicle routing problem, and in [48, 116, 242], where power-management is explicitly appended to the problem formulation. In [203, 266], the robot team is split into "task robots," which are in charge of executing the primary tasks, and "delivery robots," which supply the task robots with the required energy resources.

map the robot state to its position in $\mathscr{E}$, i.e., Robot i's location is given by $p_i = \pi(x_i) \in \mathscr{E}$, where $x_i \in \mathbb{R}^n$.

As different locations in the environment afford different recharging prospects for the robots, this needs to be characterized as well. For instance, solar-powered robots can only recharge in sunny spots, while robots that rely on a fixed recharging infrastructure can only recharge at designated charging or docking stations. An additional caveat that must be considered is that the availability of solar energy at a given location is not stationary, nor would a dedicated refueling vehicle necessarily remain in a single location. In other words, the availability of energy is time-varying.

To capture the time-varying nature of the robot-environment interaction, consider the time-varying scalar field,

$$I : \mathscr{E} \times \mathbb{R}_+ \to [0, 1], \tag{5.2}$$

where the value the available energy resource can achieve at a given location is normalized to fall in the interval $[0, 1]$. In other words, $I(p, t) \in [0, 1]$ is the intensity (e. g., solar light intensity) associated with position $p \in \mathscr{E}$ at time t, and for technical reasons we insist on I being Lipschitz continuous in its first, and differentiable in its second argument. As robots move around in the environment, $I(\pi(x_i(t)), t) \in [0, 1]$ is the intensity at Robot i's position at time t. This gives us a general and convenient way of coupling the robot to the environment.

Next on the agenda is to understand the variations in the robots' energy levels as they move around. Following the notational convention from Chapter 4, let $E_i \in \mathbb{R}_+$ be the battery energy level associated with Robot i. A number of models have been proposed at different levels of complexity for describing the dynamics of batteries and other energy storage devices, and we introduced an example of an overly simplistic such model in Equation 4.24.

To arrive at a formulation that is simple enough to be mathematically tractable, yet rich enough to be realistic and meaningful, we follow the general expression from [104, 304, 303], which has been empirically shown to capture the energy discharge dynamics associated with a number of robotic platforms. Let the charging and discharging dynamics of the battery be given by

$$\dot{E}_i = F(x_i, E_i, t) = k\left(w(x_i, E_i, t) - E_i\right), \tag{5.3}$$

where $k > 0$ is a charge/discharge gain, and the energy drain is given by

$$w(x_i, E_i, t) = \frac{1}{1 + \frac{1-E_i}{E_i} e^{-\lambda(I(\pi(x_i),t)-I_c)}}. \tag{5.4}$$

Here, $\lambda > 0$ and $I_c \in (0, 1)$ are two scalar values that capture the particulars of the energy modality [104, 304], and $I(\pi(x_i), t) \in [0, 1]$ is the previously introduced, time-varying scalar-valued function in Equation 5.2.

To see this model in action, the performance of the exponential charging-discharging dynamics in Equation 5.3 is considered on an actual, solar-powered robot over a 24-hour experiment, as per [303]. The result is detailed in Figure 5.1, where the top figure shows the results from collecting solar

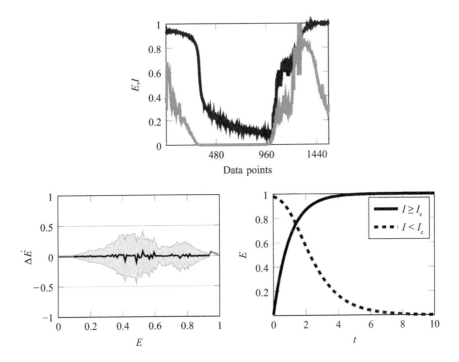

Figure 5.1: Empirical validation of the energy model in Equation 5.3 using data collected during a long-term experiment with a solar-powered robot. Top: Data collected during the course of a 24-hour experiment. E and I are the measured battery energy and solar light intensity, respectively. Bottom Left: Comparison between measured and predicted values of the energy dynamics. Here, $\Delta \dot{E}$ is the difference between the measured $\dot{E}$ and its predicted value, based on Equation 5.3. The mean of $\Delta \dot{E}$ is shown as a thick line, whereas the shaded area represents the region of one standard deviation from the mean value. Bottom Right: Simulated battery charging and discharging dynamics. A comparison with the top figure reinforces the fact that the model in Equation 5.3 is able to reproduce the dynamics of a real battery in a reasonably faithful manner. Source: Fig. 1 in Notomista et al. (2021).

panel data during the course of the experiment, including the measured battery energy and solar light intensity. In the bottom left figure, the difference, $\Delta \dot{E}$, is shown between the measured $\dot{E}$ and its predicted value using the energy model in Equation 5.3. The bottom right figure displays the simulated battery charge and discharge curves obtained using Equation 5.3 for the cases when $I > I_c$ and $I < I_c$, respectively. What this figure underscores is that the model is indeed able to capture the exponential charging and discharging dynamics of real batteries.

A final observation about this particular choice of energy dynamics is that the lack of an explicit dependence on the control input, u_i, is a purposeful choice rather than an oversight. The energy dynamics are intended to reflect a conservative approximation, thereby increasing the robustness of the persistification strategy. In fact, the rate of charge and discharge of the battery, obtained when $I(\pi(x_i), t) = 1$ and $I(\pi(x_i), t) = 0$, respectively, are designed to be the rates obtained when the robot input, u_i, attains its maximum value (when the actuators are absorbing maximum power), i.e., when the battery discharges most rapidly or charges most slowly. Under this approximation, the actual discharge rate will be lower than what the model prescribes, thereby providing robustness to the forthcoming persistification strategies against unmodeled factors in the environment and/or the battery dynamics.

These models are rich enough to capture quite a large set of situations. For example, in the case where the intensity, $I(\pi(x_i), t)$, represents a measure of the solar light intensity at the position $\pi(x_i)$ at time t, $F(x_i, E_i, t)$ describes, at least qualitatively, the energy dynamics of a solar rechargeable battery in that

- $\dot{E}_i < 0$, i.e., the battery is discharging when $I(\pi(x_i), t) < I_c$;
- $\dot{E}_i > 0$, i.e., the battery is charging when $I(\pi(x_i), t) > I_c$; and
- $\dot{E}_i = 0$ when $I(\pi(x_i), t) = I_c$, i.e., the generated energy is equal to the energy required by the robot at time t.

Figure 5.2 shows an example of how this coupling between energy dynamics and environmental resources works. In the figure, the surface plot of the function I, at a given time instant t, is displayed in grayscale (black to white for values going from 0 to 1). Below the surface plot, the contour plot of I elucidates the level curves where $I(\cdot, t) = I_c$. Inside the regions bounded by the bold curves, characterized by $I(\cdot, t) > I_c$, we have that $\dot{E}_i > 0$, i.e., the robots can charge their batteries at those locations.

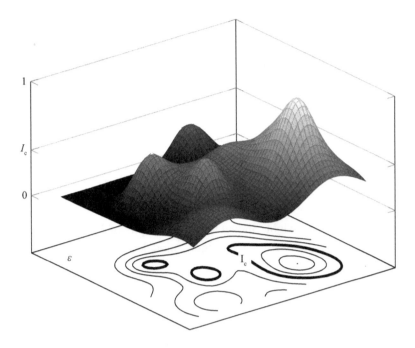

Figure 5.2: Example of the function $I(\cdot, t)$ over an environment $\mathscr{E}$ at a given time instant. In the interior of the bold level curves, it holds that $\dot{E}_i > 0$, i.e., here robots can recharge their batteries effectively. Outside, on the other hand, the energy levels are decreasing, i.e., $\dot{E}_i < 0$. Source: Fig. 2 in Notomista et al. (2021).

As solar energy is a readily available resource, with an ubiquitous presence in most natural environments, it will serve as the primary energy source under consideration. However, another common scenario that must be captured is one where the robots have to go to a dedicated location to recharge/refuel. Such "lumped" sources of energy, e.g., charging stations, can also be modeled using Equation 5.2. For instance, bump-like functions [411] placed at the locations of the charging stations can be employed to this end in order to obtain the desired charging characteristics, as depicted in Figure 5.3.

Equipped with both an energy model and a characterization of how robots can replenish their energy caches by interacting with the environment, we are now ready to move on to the next item on the persistification agenda. What is needed is an understanding of how one can use these models in order to ensure that the robots can execute their tasks indefinitely, while never running out of battery.

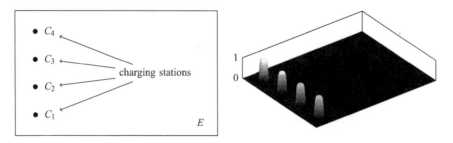

Figure 5.3: An example of lumped sources of energy (charging stations), modeled by a suitable $I(\cdot, t)$ function. Left: A rectangular environment, $\mathcal{E}$, is shown and the positions of four charging stations, denoted by C_1 to C_4, are depicted as black dots. Right: Shown is the surface plot of $I(\cdot, t)$, corresponding to the charging stations in the left figure, modeled by means of bump-like functions [411].

5.1.2 Task Persistification

In the introductory chapter, a number of key principles were established for long-duration deployments. One of these principles prescribed moving away from specific missions, and avoiding solutions that were overly targeted at the particulars of the primary tasks. As such, in order to achieve persistent environmental monitoring, we first need to understand how to persistify general robotic tasks.

As in the previous chapter, consider a collection of N robots, with the dynamics of Robot i being given by the control-affine dynamical system in Equation 4.6, $\dot{x}_i = f(x_i) + g(x_i)u_i$, $i \in [N]$, where $x_i \in \mathbb{R}^n$ is the robot's state, $u_i \in \mathbb{R}^m$ its control input, and f and g are two locally Lipschitz continuous vector fields. Combining this with the energy dynamics in Equation 5.3 produces the compound model

$$\dot{x}_i = f(x_i) + g(x_i)u_i$$
$$\dot{E}_i = F(x_i, E_i, t). \tag{5.5}$$

Indicating the augmented state of Robot i by $\mathbf{x}_i = [x_i^T, E_i]^T$, the full-fledged robot model retains the control-affine structure in that

$$\dot{\mathbf{x}}_i = \hat{f}(\mathbf{x}_i, t) + \hat{g}(\mathbf{x}_i)u_i, \tag{5.6}$$

where

$$\hat{f}(\mathbf{x}_i, t) = \begin{bmatrix} f(x_i) \\ F(x_i, E_i, t) \end{bmatrix} \quad \text{and} \quad \hat{g}(\mathbf{x}_i) = \begin{bmatrix} g(x_i) \\ 0 \end{bmatrix}. \tag{5.7}$$

To help keep the notation compact, we will use

$$\mathbf{x} = \begin{bmatrix} \mathbf{x}_1 \\ \vdots \\ \mathbf{x}_N \end{bmatrix} \in \mathbb{R}^{N(n+1)} \quad \text{and} \quad u = \begin{bmatrix} u_1 \\ \vdots \\ u_N \end{bmatrix} \in \mathbb{R}^{Nm}$$

to represent the joint states and inputs of the N robots. Furthermore, let the primary task, i.e., the task to be persistified, be represented by the nominal control input, $u_{nom,i}(\mathbf{x}, t)$, $i \in [N]$, or, collectively, as

$$u_{nom} : \mathbb{R}^{N(n+1)} \times \mathbb{R}_+ \to \mathbb{R}^{Nm}. \tag{5.8}$$

Note that this choice of task representation is quite general insofar as it encompasses both feedback controllers, through the dependence of u_{nom} on the state $\mathbf{x}$, and controllers generated by high-level planning strategies or external user inputs, through u_{nom}'s dependence on time, t.

Returning to the constrained optimization formalism introduced in Chapter 4, what we aim to do is find control inputs that render the safe set (*never run out of energy*) forward invariant. The safe set is in this case given by

$$S = \left\{ \mathbf{x} \in \mathbb{R}^{N(n+1)} \mid E_i \in [E_{\min}, E_{\max}], \ i \in [N] \right\}, \tag{5.9}$$

where $E_{\min}$ and $E_{\max}$ are the minimum and maximum energy values, respectively, between which the energy of the robots should be confined.[3] These energy bounds satisfy $1 \geq E_{\max} > E_{\min} > 0$. And, if $E_i \in [E_{\min}, E_{\max}]$ can be ensured for all times, then we say that the task has been *persistified* relative to a suitable measure of task completion. Following the cost formalism introduced in Equation 4.19, let $\mathscr{C}(u, u_{nom}, \mathbf{x}, t)$ be a general, instantaneous cost that measures how well (or poorly) the task encoded through u_{nom} is being executed.

---

[3] The reason why this energy constraint is simpler than the one in Equation 4.30, from the previous chapter, stems from the fact that the energy dynamics in Equation 5.5 depends on x_i, which in turn depends on u_i. As such, it is now possible to influence the energy state, E_i, indirectly through the control input, u_i, which was not the case before.

It is now tempting to, once again, formulate task persistification as

$$u^\star = \text{argmin}_{u \in \mathbb{R}^{Nm}} \, \mathscr{C}(u, u_{nom}, \mathbf{x}, t),$$

subject to

$$E_i(t) \in [E_{\min}, \, E_{\max}], \quad \forall i \in [N].$$

Unfortunately, this formulation suffers from the same problems as Equation 4.1 in Chapter 4 in that we are asking for a minimizing control input subject to an instantaneous constraint that does *not* involve the input. And, as already discussed, this makes it an ill-posed problem. Luckily for us, we now know how to remedy this through the CBF machinery.

Based on the observation that the energy level, E_i, cannot be negative, the two different constraints, $E_i \geq E_{\min}$ and $E_i \leq E_{\max}$, can be turned into a single, scalar constraint,

$$h_{energy}(\mathbf{x}_i) = (E_{\max} - E_i)(E_i - E_{\min}) \geq 0, \tag{5.10}$$

with the corresponding CBF constraint, $S_{h_{energy}}[\mathbf{x}_i; u_i] \geq 0$, ensuring forward invariance of the safe set, S, in Equation 5.9.

The task persistification problem thus becomes:

Problem 5.1 Given the instantaneous performance cost, $\mathscr{C}(u, u_{nom}, \mathbf{x}, t)$, find the minimizing $u^\star$ to the constrained optimization problem

$$u^\star = \text{argmin}_{u \in \mathbb{R}^{Nm}} \, \mathscr{C}(u, u_{nom}, \mathbf{x}, t),$$

subject to

$$S_{h_{energy}}[\mathbf{x}_i; u_i] \geq 0, \quad \forall i \in [N],$$

given the constraint, h_{energy}, in Equation 5.10.

If this problem is solved, we say that the task has been $\mathscr{C}$-*persistified*.

Definition 5.1 — Task Persistification. A task encoded through the nominal input, u_{nom}, is $\mathscr{C}$-persistified by the minimizer to the problem in Problem 5.1.

A particularly pertinent choice of cost is the L_2 norm, $\|u - u_{nom}\|^2$, and we will simply say that the task has been *persistified* by the minimizer to the problem

$$u^\star = \operatorname{argmin}_{u \in \mathbb{R}^{Nm}} \|u - u_{nom}\|^2,$$

subject to

$$S_{h_{energy}}[\mathbf{x}_i; u_i] \geq 0, \ \forall i = 1, \ldots, N.$$

The next step towards achieving task persistification is to unleash the full force of the CBF machinery. But, the time derivative of the CBF in Equation 5.10 is given by

$$\dot{h}_{energy}(\mathbf{x}_i) = (E_{max} + E_{min} - 2E_i)F(x_i, E_i, t), \tag{5.11}$$

which exhibits two previously unseen peculiarities, namely that *(i)* it is a function of time t, and perhaps more disturbing, *(ii)* it is a function of $\mathbf{x}_i$ but *not* of u_i. As such, we still do not have a way of ensuring that the constraint is satisfied unless we somehow introduce u_i into the mix. The way to accomplish this is to take at least one more derivative of h_{energy} to get to an expression involving the control input, which is necessary in order to be able to find a control signal that renders the safe set forward invariant. This, in turn, means that we need to understand how to design CBF-based controllers where multiple time-derivatives are needed, i.e., for "outputs" with high *relative degree*.

5.2   Variations on the CBF Theme

The outcome of the previous discussion is that the CBF framework needs to be augmented somewhat in order to handle time-varying CBFs as well as CBFs with high relative degree. As the former of the two is less involved, we start with the thornier question of how to ensure forward invariance of the safe set when the control input does not show up in the first derivative of the CBF.

5.2.1   High Relative Degree Barrier Functions

The reason for having to investigate how to handle high relative degree situations is due to the energy dynamics in Equation 5.3. As a consequence of this modeling choice, if the output to the system is taken to be the CBF in

Equation 5.10, the relative degree is greater than 1 as the energy dynamics do not depend on u_i. As discussed, this choice was made for robustness reasons. But the gain in robustness comes at the price of increasing the relative degree of the CBF.

To get started, we first need to formally establish what is meant by the relative degree of an input-output system. Given a control-affine system, $\dot{x} = f(x) + g(x)u$, with f and g being sufficiently smooth on a domain $\mathcal{D}$, we can associate an output, $y = h(x)$, with this system, where h is some sufficiently smooth function. This output could be an actual, measured output. Or it could just be some particular aspect of the state that we care about for some reason, like a compact description of what "robot survival" might entail.

The time derivative of the output is

$$\dot{y} = \frac{\partial h^T}{\partial x}\dot{x} = \frac{\partial h^T}{\partial x}f(x) + \frac{\partial h^T}{\partial x}g(x)u = L_f h(x) + L_g h(x)u, \qquad (5.12)$$

where we have used the Lie-derivative notation introduced in Equation 4.13. If $L_g h(x) \neq 0$, then $\dot{y}$ depends on u, and we say that the system has relative degree 1 (at x). But if this is not the case, i.e., if $L_g h(x) = 0$, we have that

$$\dot{y} = L_f h(x)$$

and, as a result, we need to take additional derivatives,

$$\ddot{y} = \frac{\partial L_f h^T}{\partial x}\dot{x} = \frac{\partial L_f h^T}{\partial x}f(x) + \frac{\partial L_f h^T}{\partial x}g(x)u. \qquad (5.13)$$

Based on the notational convention [366] that

$$\frac{\partial L_f h^T}{\partial x}f(x) = L_f^2 h(x) \quad \text{and} \quad \frac{\partial L_f h^T}{\partial x}g(x) = L_g L_f h(x),$$

we can rewrite $\ddot{y}$ more compactly as

$$\ddot{y} = L_f^2 h(x) + L_g L_f h(x)u. \qquad (5.14)$$

Now, if $L_g L_f h(x) \neq 0$, then u shows up in $\ddot{y}$ and the relative degree is 2. If it is equal to zero, we have to keep going until u shows up.

> **Definition 5.2 — Relative Degree.** The system $\dot{x} = f(x) + g(x)u$, with output $y = h(x)$, has relative degree, $\rho \in \mathbb{Z}_+$, at x_0, if

$$L_g L_f^k h(x) = 0, \ \forall x \text{ in a neighborhood of } x_0, \text{ and for all } k \leq \rho - 2 \tag{5.15}$$

$$L_g L_f^{\rho-1} h(x_0) \neq 0. \tag{5.16}$$

Luckily for us, CBFs with high relative degree are actually not all that hard to handle, as demonstrated in [298, 304]. Consider an example of a CBF candidate, $h_1 : \mathbb{R}^n \to \mathbb{R}$, with relative degree 2 (meaning that $L_g h_1(x) = 0$ and $L_g L_f h_1(x) \neq 0$) that defines the super-level set $S_1 = \{x \in \mathbb{R}^n \mid h_1(x) \geq 0\}$. To establish forward invariance of this set, we need h_1 to be a CBF, for which the following condition must hold,

$$L_f h_1(x) + L_g h_1(x)u + \alpha_1(h_1(x)) \geq 0,$$

where α_1 is some continuously differentiable, extended class $\mathscr{K}_\infty$-function. But, as the relative degree is 2, $L_g h_1(x) = 0$, and the inequality condition is actually

$$L_f h_1(x) + \alpha_1(h_1(x)) \geq 0. \tag{5.17}$$

In essence, what we have done is gone from the condition that

$$h_1 \geq 0$$

to a new condition that states that

$$L_f h_1(x) + \alpha_1(h_1(x)) \geq 0.$$

As a result, we have traded one condition for another, and we can thus proceed to define a second CBF candidate, h_2, as

$$h_2(x) = L_f h_1(x) + \alpha_1(h_1(x)), \tag{5.18}$$

whose super-level set is $S_2 = \{x \in \mathbb{R}^n \mid h_2(x) \geq 0\}$. The time derivative of h_2 is, in turn, given by

$$\dot{h}_2 = L_f^2 h_1(x) + L_g L_f h_1(x)u + \frac{\partial \alpha_1}{\partial h_1} L_f h_1(x), \tag{5.19}$$

where we, once again, used the fact that $L_g h_1(x) = 0$. Since the relative degree was 2, we have that $L_g L_f h_1(x) \neq 0$ and, as a result, u shows up explicitly in Equation 5.19. This gives us something to latch on to from a constrained optimization vantage point.

In light of these derivations, if there exists a locally Lipschitz, extended class $\mathcal{K}_\infty$-function, α_2, such that

$$\sup_{u \in \mathbb{R}^m} \left\{ L_f^2 h_1(x) + L_g L_f h_1(x) u + \frac{\partial \alpha_1}{\partial h_1} L_f h_1(x) + \alpha_2(h_2(x)) \right\} \geq 0,$$
(5.20)

then h_2 is a valid CBF. And this means that S_2 is rendered forward invariant, i.e., that the condition in Equation 5.17 also holds for that choice of input. But, that means that S_1 is also rendered forward invariant by the same choice of input, i.e., that h_1, in turn, is a valid CBF as well, As a direct consequence, the question of how to handle CBFs with relative degree 2 has been resolved.

The technique for CBFs with relative degree 2 can be generalized through a cascade of CBFs to the following (complicated looking but quite straightforward) theorem, found in a slightly different form in [298, 302],

---

Theorem 5.1 Given the control-affine system, $\dot{x} = f(x) + g(x)u$, together with a sufficiently smooth CBF candidate, $h_1(x)$, with relative degree $\rho > 1$, the CBF, $h_\rho(x)$, can be recursively evaluated from $h_1(x)$ as follows,

$$h_{q+1}(x) = \dot{h}_q(x) + \alpha_q(h_q(x)), \quad 1 \leq q < \rho,$$
(5.21)

with α_q, $q = 1, \ldots, \rho - 1$, being a sequence of continuously differentiable, extended class $\mathcal{K}_\infty$-functions. Let $K_\rho(x)$ be given by

$$K_\rho(x) = \left\{ u \; \middle| \; L_f^\rho h_1(x) + L_g L_f^{\rho-1} h_1(x) u \right.$$

$$\left. + \sum_{i=1}^{\rho-1} \sum_{J \in \binom{\rho-1}{i}} \prod_{j \in J} \frac{\partial \alpha_j}{\partial h_j} L_f^{\rho-i} h_1(x) + \alpha_\rho(h_\rho(x)) \geq 0 \right\},$$
(5.22)

where $\binom{\rho-1}{i}$ is the set of i-combinations from the set $\{1, \ldots, \rho - 1\} \subset \mathbb{N}$, and α_ρ is a locally Lipschitz, extended class $\mathcal{K}_\infty$-function. Then, any Lipschitz continuous controller, $u \in K_\rho(x)$, will render the set $S_1 = \{x \in \mathbb{R}^n \mid h_1(x) \geq 0\}$ forward invariant.

Theorem 5.1 provides exactly the structure required to handle high relative degree barrier functions. And, as seen, this is needed because the constraint in Equation 5.10 has relative degree greater than 1. (In fact, it has relative degree 2.) The additional complication is that the energy dynamics in Equation 5.3 depends explicitly on time, i.e., it is not time-invariant. That this has to be the case follows immediately from the simple observation that sometimes the sun is up. And sometimes it is not. As such, any recharging modality relying on solar energy is bound to possess a time-varying aspect. This is the topic of the next section.

5.2.2 Time Varying Barrier Functions

In order to solve the persistification problem, the next order of business is to extend the CBF formalism to the situation in which the function h depends explicitly on time. In other words, we wish to ensure the forward invariance of the (time-varying) safe set, $S(t) \subset \mathbb{R}^n$, defined by the super-level set of the function $h : \mathbb{R}^n \times \mathbb{R}_+ \to \mathbb{R}$, as

$$S(t) = \left\{ x \in \mathbb{R}^n \mid h(x,t) \geq 0 \right\}. \tag{5.23}$$

With this slight change in perspective, the definition of a CBF, given in [436], extends to the time-varying case in the following manner:

> **Definition 5.3 — Time-Varying CBF.** Given a control-affine dynamical system and a set, $S(t)$, defined in Equation 5.23, the function h is a time-varying CBF, defined over a domain $\mathscr{D} \times \mathbb{R}_+$, with $S(t) \subseteq \mathscr{D} \subset \mathbb{R}^n$, $\forall t \geq t_0$, if there exists a locally Lipschitz continuous, extended class $\mathscr{K}_\infty$-function α, such that, $\forall x \in \mathscr{D}$,
>
> $$\sup_{u \in \mathbb{R}^m} \left\{ \frac{\partial h}{\partial t} + L_f h(x,t) + L_g h(x,t) u + \alpha(h(x,t)) \right\} \geq 0. \tag{5.24}$$

Analogously to the time-invariant case, one can proceed to define the set of admissible control inputs,

$$K(x,t) = \left\{ u \in \mathbb{R}^m \, \middle| \, \frac{\partial h}{\partial t} + L_f h(x,t) + L_g h(x,t) u + \alpha(h(x,t)) \geq 0 \right\}. \tag{5.25}$$

The time-varying extension states that $S(t)$, as defined in Equation 5.23, is rendered forward invariant by the application of a control input, $u \in K(x, t)$, which we assert as a lemma,[4] found in [302],

---

Lemma 5.1 Given a set, $S(t)$, defined in Equation 5.23. If h is a time-varying CBF on $\mathscr{D} \times \mathbb{R}_+$, then any controller, $u \in K(x, t)$, that is locally Lipschitz continuous in x and piecewise continuous in t, and where $K(x, t)$ is given in Equation 5.25, will render $S(t)$ forward invariant.

5.2.3 Solving the Persistification Problem

In light of the updated CBF machinery, we now have all the necessary tools at our disposal to properly manage the task persistification problem, as understood in Definition 5.1, through the constrained optimization formulation in Problem 5.1. To this end, we start the high relative degree CBF cascade by setting $h_1 = h_{energy}$ from Equation 5.10. As h_1 does not explicitly depend on t, its time derivative is

$$
\begin{aligned}
\dot{h}_1(\mathbf{x}_i, t) &= L_{\hat{f}} h_1(\mathbf{x}_i, t) + L_{\hat{g}} h_1(\mathbf{x}_i) u_i \\
&= \left[\frac{\partial h_1^T}{\partial x_i} \frac{\partial h_i}{\partial E_i} \right] \hat{f}(\mathbf{x}_i, t) + \left[\frac{\partial h_1^T}{\partial x_i} \frac{\partial h_1}{\partial E_i} \right] \hat{g}(\mathbf{x}_i) u_i \\
&= (E_{max} + E_{min} - 2E_i) F(x_i, E_i, t),
\end{aligned}
\tag{5.26}
$$

which does not depend on u_i, as the relative degree of $h_1(\mathbf{x}_i)$ is 2. But, it does depend on t. Therefore, we define the next CBF, $h_2(\mathbf{x}_i, t)$, as in Equation 5.18,

$$
h_2(\mathbf{x}_i, t) = \dot{h}_1(\mathbf{x}_i, t) + \gamma_1 h_1(\mathbf{x}_i),
\tag{5.27}
$$

where we, for simplicity, have chosen $\alpha_1(z)$ to be the linear function $\alpha_1(z) = \gamma_1 z$, for some $\gamma_1 > 0$.

The only thing remaining is to combine Theorem 5.1 with Lemma 5.1 to define the set of control inputs that will render the set $S_1 = \{\mathbf{x}_i \in \mathbb{R}^{n+1} \mid h_1(\mathbf{x}_i) \geq 0\} = \{(x_i, E_i) \in \mathbb{R}^n \times \mathbb{R}_+ \mid E_{min} \leq E_i \leq E_{max}\}$

---

[4] In case $\frac{\partial h}{\partial t} = 0$ and $L_g h(x, t) = 0$, we are not able to ensure the existence of a control input such that the condition in Equation 5.24 holds, which is a pillar on which Lemma 5.1 rests. This case can be tackled by once again making use of a *cascade* of control barrier functions, combined with the construction in Theorem 5.1.

forward invariant. This is achieved by the following set

$$K_2(\mathbf{x}_i, t) = \left\{ u_i \in \mathbb{R}^m \;\middle|\; \frac{\partial h_2}{\partial t} + L_{\hat{f}}^2 h_1(\mathbf{x}_i, t) + L_{\hat{g}} L_{\hat{f}} h_1(\mathbf{x}_i) u_i \right.$$

$$\left. + \gamma_1 L_{\hat{f}} h_1(\mathbf{x}_i, t) + \gamma_2 h_2(\mathbf{x}_i, t) \geq 0 \right\}, \qquad (5.28)$$

where $\alpha_2(z) = \gamma_2 z$, with $\gamma_2 > 0$.

Gathering all the pieces together, we conclude this section by formally stating the fact that task persistification, as defined in Definition 5.1, has been successfully achieved:

Theorem 5.2 If $u^\star = [u_1^{\star T}, \ldots, u_N^{\star T}]^T$ is the solution to the constrained optimization problem

$$u^\star = \operatorname{argmin}_u \|u - u_{nom}\|^2,$$

subject to

$$u_i^\star \in K_2(\mathbf{x}_i, t), \ \forall i \in [N],$$

with $K_2(\mathbf{x}_i, t)$ given in Equation 5.28, then $u^\star$ persistifies the task represented by the nominal control input, u_{nom}.

5.3 Environmental Monitoring

To contextualize the task persistification framework, we now consider two robotic tasks whose persistent application is particularly relevant to robots deployed in an environment over long time-scales. These two tasks are variations on a larger application theme, namely environmental monitoring, where robots are to be deployed to collect information about an area. In particular, the tasks under consideration are environmental exploration and coverage. Both of these tasks are typically required to be executed for a long period of time. In the case of environmental exploration, the long execution time is due to the size (think Mars) and/or the dynamic nature of the environment [42, 158]. Regarding environmental coverage, the time-scales of the observed phenomena are the primary factors in determining the duration of the deployment [124]. For instance, agricultural robots may need to be present in the farm field for the entire growing cycle, i.e., at a time-scale dictated by the plants themselves [23].

5.3.1 Exploration

Suppose that the robots are tasked with moving around in an environment in order to gather as much information as possible. As before, let the environment be given by $\mathcal{E}$. Moreover, assume that "information" is available over $\mathcal{E}$, and that this information is spread spatially according to the distribution $\phi : \mathcal{E} \mapsto \mathbb{R}_+$, where a higher ϕ-value means that more information is available at that location, akin to the density function in the locational cost in Equation 2.19.

There are many different approaches one can take to solving the exploration task, and a number of qualitatively different solutions have been proposed, e.g., [42, 158]. This plethora of options does not matter for the purpose of persistification as they simply result in different nominal control inputs, u_{nom}, i.e., it ultimately does not matter for the exposition how the nominal controller is originated and produced. In this chapter, we choose to follow the route established in [279], where the nominal control input is the outcome from a trajectory optimization problem. In particular, the optimal trajectories (and, subsequently, inputs) are obtained by making them as *ergodic* as possible, i.e., by producing trajectories that traverse the environment in such a way that the time spent in any one region is proportional to the total amount of information density present in that region [333]. In [265], an ergodic metric was introduced over trajectories and a computational framework was proposed for finding the optimal trajectories that maximize this metric.

Since, at the end of the day, the persistification strategy does not depend on the manner in which u_{nom} is obtained, we do not dwell here on the particulars of how to find such ergodic trajectories, and instead refer to [265] for the details. Implementation details aside we thus, at the current time t_c and over a time horizon T, have access to a nominal trajectory $\hat{x}_{nom,i}(t)$ for $t \in [t_c, t_c + T]$, for each of the robots, and a nominal controller, $u_{nom,i}$, that produces this trajectory—either exactly or approximately, as was done in [302]. At each instant of time, the constrained optimization problem for persistifying the exploration task can subsequently be solved. This can moreover be achieved in real time as the optimization problem is, as already seen in Chapter 4 (Equation 4.20), a computationally unproblematic quadratic programming problem.

The persistent environmental exploration task is implemented and tested in a simulated environment. In the experiment, a single, planar robot is given the task of exploring the environment, $\mathcal{E}$, over which the spatial distribution of information has been defined. This information is distributed according to the static, Gaussian density function

$$\phi(x) = e^{-\frac{\|x - x_0\|^2}{\sigma^2}} \in \mathbb{R}_+, \tag{5.29}$$

where $x \in \mathbb{R}^2$, and for the purpose of the experiment, $x_0 = [0, 0]^T$ and $\sigma^2 = 0.1$. The time-varying environment field, I, is modeled as a mixture of time-varying Gaussians,

$$I(x, t) = e^{-\|x - M_1(t)x_c\|^2} + e^{-\|x - M_2(t)x_c\|^2}, \tag{5.30}$$

where $x_c = [1, 1]^T$, and

$$M_1(t) = \begin{bmatrix} -1 & 0 \\ 0 & \sin(2t) \end{bmatrix}, \quad M_2(t) = \begin{bmatrix} \sin(2t) & 0 \\ 0 & 1 \end{bmatrix}. \tag{5.31}$$

In the case of robots that are able to exploit solar power to recharge their batteries, this choice of I simulates the sunlight intensity that is characterized by a periodic expression over a spatially fixed environment.[5] (The other values required to model $\dot{E}_i$ in Equation 5.3 are set to $I_c = 0.85$ and $\lambda = 3$.)

In the experiment, the output from the ergodic trajectory optimization process results in a nominal position, predicted a short distance into the future, for the robot to move towards, and the corresponding nominal control input is one that simply drives the robot towards that location. The nominal position is depicted as a black square in Figure 5.4, while the actual position as a black circle. (Nominal and executed trajectories are also shown, represented by a thick solid line and a thick dashed line, respectively.) Figure 5.4 displays a sequence of snapshots taken during the course of the environmental exploration experiment. The contour plot of the information density function, ϕ, is depicted with thin solid lines, and the nominal controller is driving the robot in such a way so as to maximize the information gathered about the environment. In addition to the information density, the energy intensity field, I, must

---

[5]This almost circadian aspect of the prevalence of solar energy is explicitly leveraged for slow-paced robots in [307].

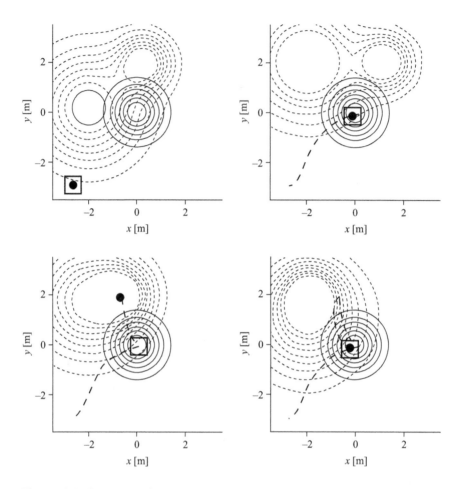

Figure 5.4: Snapshots from an environmental exploration experiment. The contour plots of the information distribution function, ϕ, and the environmental field, I, are depicted as thin solid lines and thin dashed lines, respectively. The nominal position, specified by the ergodic trajectory optimization process, is represented as a square, while the actual position of the robot is depicted as a black circle. The nominal and actual trajectories are displayed as thick solid lines, and thick dashed lines, respectively. In order to persistently explore the environment, the robot follows the nominal input as long as its energy level is sufficiently high (Top Left and Top Right). When its battery has depleted to a critical level, the robot moves towards regions of the environment (Bottom Left), where the value of the time-varying field, I, is such that its energy starts to increase. Once sufficiently recharged, the robot returns to the nominal position (Bottom Right). Source: Fig. 5 in Notomista et al. (2021).

also be taken into account, and it is represented by the thin, dashed lines in the figure.

The most salient aspect of the sequence of simulation snapshots in Figure 5.4 is when the robot starts deviating from the ergodic trajectory towards the end of the sequence. Once it becomes energetically beneficial to no longer follow the ergodic trajectory, the robot instead takes a detour towards areas with higher energy densities. When the batteries are sufficiently replenished, the robot returns to where it left off, and continues following the ergodic trajectory. The beauty of this approach is that no explicit tradeoff is needed in terms of exploration versus recharging. Instead, the persistification just happens organically, as a consequence of the constrained optimization framework.

The exploratory quality of the persistified environmental exploration strategy can be surmised from Figure 5.5. There, the true spatial information density distribution, ϕ, is given in the left figure. As the robot moves around in the environment, it gathers information about ϕ and subsequently builds up its own estimated version of the information available in the environment. As is seen in the middle and right figures, this estimated version is getting increasingly accurate as more information is amassed [302].

The real punchline with the environmental exploration experiment is the persistification of the task, rather than the task itself. In fact, this persistification concept was depicted already in Figure 3.9, which shows the energy level of the robot during the exploration experiment discussed here. As expected, the battery level remains within the specified bounds throughout the deployment. The corollary to this is that the robot can indeed perform the environmental exploration task indefinitely.

5.3.2 Coverage

Related to environmental exploration is environmental coverage, where robots are to spread out across an area for the purpose of detecting events or other phenomena of interest. In contrast to the exploration problem, coverage is an inherently multi-robot affair, as the former can be achieved by a single robot moving purposefully through the environment, gathering information, while the latter inevitably involves having multiple vantage points (i.e., robots) to cover the area sufficiently well. In fact, for the coverage problem, the robots are expected to use their sensors to keep track of events happening in their area of responsibility, also known as their *regions of dominance* to use the terminology from Chapter 2. An additional difference between coverage and

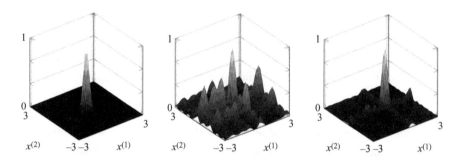

Figure 5.5: Comparison between the true, spatial probability density function (Left) and the estimated probability density function. The estimates (Middle and Right) are obtained during the persistified environmental exploration process. Over time, a better estimate is produced as the robot is able to gather more information as it moves around in the environment.

exploration is that when the environment is static and no refueling is required, once the robots are properly distributed, the solution to the coverage problem is for the robots to remain stationary. In other words, coverage is not inherently dynamic the way exploration is.

In light of the previous paragraph, the task of environmental coverage can be framed as a sensor placement problem [314]. As before, given an environment, $\mathcal{E} \subset \mathbb{R}^p$, with $p = 2$ or 3, we can follow the recipe from Chapter 2 and associate the quality (actually, lack thereof), $Q(\|\pi(x_i) - q\|)$, with the measurement of a point $q \in \mathcal{E}$, taken by a robot located at position $\pi(x_i)$. Analogous to what is found in territorial animal behaviors [308, 357, 370], encoded through the resource integrals in Chapter 3, we let Robot i be responsible for everything inside a particular subset of the environment, $W_i \subseteq \mathcal{E}$. Morever, as in the previous section, we let the map $\phi : \mathcal{E} \to \mathbb{R}_+$ represent a spatial density function. This can be interpreted as a measure of the information spread over the environment, $\mathcal{E}$, or, if properly normalized, as the probability density associated with the likelihood of an event taking place at a given location.

Gathering everything together, we obtain the familiar-looking cost associated with the placement of Robot i, given the region of dominance, W_i, as

$$\int_{W_i} Q(\|\pi(x_i) - q\|)\phi(q)dq.$$

We now insist on having the regions of dominance correspond to a proper tessellation of the environment, $W = \{W_1, \ldots, W_N\}$, i.e.,

$$\bigcup_{i=1}^{N} W_i = \mathscr{E}, \quad W_i^{\circ} \cap W_j^{\circ} = \emptyset, \ \forall i \neq j, \tag{5.32}$$

where W_i° is the interior of the set W_i, which gives the locational cost [95]

$$\mathscr{H}(\mathbf{x}, W) = \sum_{i=1}^{N} \int_{W_i} Q(\|\pi(x_i) - q\|)\phi(q)dq. \tag{5.33}$$

Proceeding as in [95], $\mathscr{H}(\mathbf{x}, W)$ is minimized (locally), if W is a Voronoi tessellation in the sense that $W_i = V_i(\mathbf{x})$, with $V_i(\mathbf{x})$ being the Voronoi cell consisting of all points in $\mathscr{E}$ closer to Robot i than to any other robots, i.e.,

$$V_i(\mathbf{x}) = \{q \in \mathscr{E} \mid \|\pi(x_i) - q\| \leq \|\pi(x_j) - q\| \ \forall i \neq j\}. \tag{5.34}$$

Replacing W_i by $V_i(\mathbf{x})$ in Equation 5.33 produces the familiar-looking locational cost, solely dependent on the robots' states, as

$$\mathscr{H}(\mathbf{x}) = \sum_{i=1}^{N} \int_{V_i(\mathbf{x})} Q(\|\pi(x_i) - q\|)\phi(q)dq. \tag{5.35}$$

As far as the minimization of the robot locations is concerned, we once again let the robots move against the gradient of the locational cost in Equation 5.35, as was done in [276] as well as in Equation 2.20. This choice, in turn, produces the required, nominal controller for the constrained optimization problem.[6] The minimizing direction in which Robot i should nominally be moving is given by $\rho_i(\mathbf{x}) - \pi(x_i)$, where $\rho_i(\mathbf{x})$ is the centroid of the ith Voronoi cell,

$$\rho_i(\mathbf{x}) = \frac{\int_{V_i(\mathbf{x})} q\phi(q)dq}{\int_{V_i(\mathbf{x})} \phi(q)dq}, \tag{5.36}$$

as per Equation 2.23.

The persistified environmental coverage strategy is implemented and deployed on the Robotarium,[7] which is a remotely accessible swarm robotics research testbed [337]. In the experiment, a team of 7 mobile robots are tasked

---

[6]In case the map, ϕ, is time-varying, the extension presented in [229, 365] can be directly employed.

[7]The Robotarium is covered in detail in Chapter 9 as a canonical instantiation of the autonomy-on-demand concept.

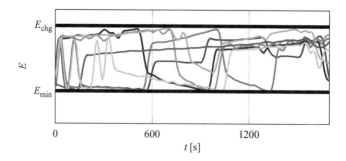

Figure 5.6: Measured battery levels of 7 robots executing a persistified coverage control experiment. As expected, the energy levels are indeed kept between the maximum and minimum levels.

with covering the Robotarium arena. And, as the Robotarium is endowed with wireless charging stations, the charging field, I, is given in Equation 5.2 by means of bump-like functions, as illustrated in Figure 5.3. To show that the persistification strategy is indeed working as expected, Figure 5.6 displays the energy stored in the batteries during the course of the experiment.

The sequence in Figure 5.7 shows four salient frames of the persistent environmental coverage experiment. The Voronoi cells are superimposed on the arena by means of an overhead projector, while the wireless charging stations are arranged along one of the edges of the testbed. Following the nominal controller, obtained by solving the coverage control problem, the robots perform sensor coverage while their energy caches are healthy. As their stores start to deplete to dangerous levels, they go to the charging stations and recharge their batteries to prevent the energy from going below $E_{\min}$.

To summarize the key takeaway from this chapter, we have shown how to take a robotic task, represented by a nominal controller, and render it persistent by ensuring that the following constraint always holds,

$$E_i \in [E_{\min}, E_{\max}], \ i \in [N],$$

which is highly reminiscent of the homeostasis constraint introduced in Chapter 3. The key to making the corresponding safe set forward invariant is the ability to turn the constraint into a single, scalar CBF,

$$h_{energy}(\mathbf{x}_i) = (E_{\max} - E_i)(E_i - E_{\min}).$$

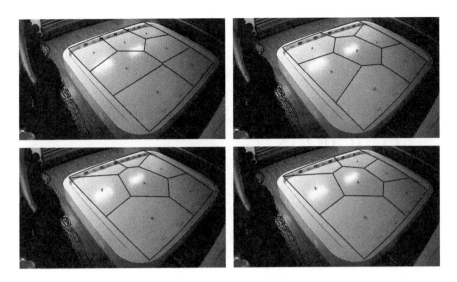

Figure 5.7: Persistent environmental coverage experiment. A team of 7 mobile robots are deployed to perform persistent sensor coverage in the Robotarium. Top Left: The robots spread out to cover the arena. Top Right: One of the robots has to go to the edge of the arena to recharge its depleted batteries. Bottom Left: A second robot goes to recharge. Bottom Right: The first recharging robot has replenished its battery sufficiently, and is returned to the nominal coverage task.

An alternative way to formulate the energy constraint would be to say that we need to ensure that

$$E_i \geq E_{\min} \text{ AND } E_i \leq E_{\max},$$

i.e., as a conjunction (AND) of two constraints. But why stop there? What if we wanted to create more elaborate constraints that involve a collection of many different constraints composed together using not only AND, but also OR, and NOT? Then we would be hard-pressed to come up with a single, scalar CBF without additional technical machinery. *In order to go from persistification to full-fledged survivability, we need the ability to produce more complicated logic expressions.* This question of *Boolean composition of CBFs* is the topic of the next chapter. It constitutes the last technical ingredient needed to get to robot ecology for the purpose of attaining long-duration autonomy.

6 Composition of Barrier Functions

We have now assembled an expressive and powerful tool for managing constraint-based control design that allows us to render safe sets forward invariant, i.e., if the system starts safe, it stays safe. Additionally, we have seen how to engage this formalism around the issue of making tasks persistent by introducing minor tweaks to the control programs in order to ensure that the tasks can be performed indefinitely, beyond a single battery charge. The enabling mechanism for achieving this feat was the association of "safety" with a scalar function, h, and letting the safe set, S, be given by the super-level set to h, in the sense that $h(x) \geq 0 \Leftrightarrow x \in S$. The scalar function, h, is a *control barrier function* (CBF), and the associated invariance constraint demands that $\dot{h} \geq -\alpha(h)$ should hold for all times, given a locally Lipschitz continuous, extended class $\mathcal{K}_\infty$-function, α, as per Chapter 4.

The impetus behind these developments is the desire to ensure robot survival across long temporal scales. But it is, to be completely honest, not entirely clear how to take something as nebulous and complex as "survival" and encode such a concept with a single, scalar function, $h_{survive}$. In fact, we have already seen two fairly involved CBFs in the previous two chapters; one focusing on collision-avoidance (Chapter 4), and one on the persistification of tasks, i.e., on ensuring that robots never get stranded away from charging stations/areas, with completely depleted batteries (Chapter 5). What we would really like for the robots to do is, of course, to both avoid collisions *and* have enough gas left in the tank, i.e., we need some way of combining multiple CBFs into a single, unified, super-CBF.

6.1 Boolean Composition

To start the discussion about how to combine together multiple CBFs, consider two sets, S_1 and S_2, with associated level-set functions, h_1 and h_2, in the sense that $h_i(x) \geq 0 \Leftrightarrow x \in S_i$, $i = 1, 2$. The corresponding CBF constraints, as per Definition 4.2, can be compactly represented as $S_{h_i}[x; u] \geq 0$, $i = 1, 2$.

What if we now would like to have both $h_1(x) \geq 0$ *and* $h_2(x) \geq 0$ simultaneously? In this case, we could add both constraints to the constrained optimization problem, i.e.,

$$\min_u \mathscr{C}_{task}(x, u, u_{nom}), \tag{6.1}$$

subject to the *two* constraints

$$\begin{aligned} S_{h_1}[x; u] &\geq 0 \\ S_{h_2}[x; u] &\geq 0, \end{aligned} \tag{6.2}$$

where, as in previous chapters, $\mathscr{C}_{task}$, encodes adherence to the nominal task that the robot is asked to perform.

But, as we add more ingredients to the survival specification, we are increasing the number of constraints in a potentially unmanageable and problematic manner. Additionally, what if we wanted $x \in S_1$ *or* $x \in S_2$ instead? In that case, we cannot simply augment the list of constraints, since that is not how the disjunctive "or" works. Instead, rather than juggling multiple constraints, what is needed is a method of producing a single, scalar CBF from the constituent CBFs, h_1 and h_2.

6.1.1 Disjunctions and Conjunctions

The takeaway from the previous discussion is that we, at the very least, need to be able to handle operations such as "and" and "or" under the CBF umbrella. From a set-theoretic standpoint, these composition operators are quite unproblematic, as seen in Figure 6.1.

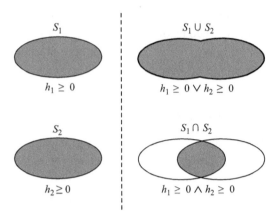

Figure 6.1: Given two super-level sets, S_1 and S_2, to the CBFs, h_1 and h_2 (Left), disjunction is given by the union operation, $S_1 \cup S_2$ (Top Right), and conjunction by the intersection, $S_1 \cap S_2$ (Bottom Right).

The *conjunction* (AND) operation, where $x \in S_1$ and $x \in S_2$, is given by the intersection of the two sets, i.e.,

$$x \in S_1 \text{ and } x \in S_2 \Leftrightarrow x \in S_1 \cap S_2. \tag{6.3}$$

Similarly, the *disjunction* (OR) is given by the union,

$$x \in S_1 \text{ or } x \in S_2 \Leftrightarrow x \in S_1 \cup S_2. \tag{6.4}$$

Unfortunately, neither of these two set operations lead, in any immediate fashion, to a CBF. For this, some additional work is needed.

Starting with the conjunction operator, we note that $h_1(x) \geq 0$ and $h_2(x) \geq 0$ as long as the smallest of the two is non-negative, as noted in [161]. In other words,

$$x \in S_1 \cap S_2 \Leftrightarrow \min\{h_1(x), h_2(x)\} \geq 0. \tag{6.5}$$

Put another way, if we let $h_{1 \wedge 2}$ be the CBF associated with $S_1 \cap S_2$, we have that

$$h_1(x) \geq 0 \wedge h_2(x) \geq 0 \Leftrightarrow h_{1 \wedge 2}(x) \geq 0,$$
$$\text{where } h_{1 \wedge 2}(x) = \min\{h_1(x), h_2(x)\}. \tag{6.6}$$

Analogously, we note that $h_1(x) \geq 0$ or $h_2(x) \geq 0$ holds when the largest of the two is non-negative, i.e.,

$$x \in S_1 \cup S_2 \iff \max\{h_1(x), h_2(x)\} \geq 0, \tag{6.7}$$

which gives the disjunction

$$h_1(x) \geq 0 \vee h_2(x) \geq 0 \iff h_{1\vee 2}(x) \geq 0,$$
$$\text{where } h_{1\vee 2}(x) = \max\{h_1(x), h_2(x)\}. \tag{6.8}$$

These two innocent-looking constructions ("and = min" and "or = max") are actually all we need to produce a rich set of constraint-based robot behaviors. The remainder of this chapter explores where this perspective will lead technically, including appearances by generalized gradients and set-valued Lie derivatives, connectivity maintenance constraints, and non-smooth CBFs.

6.1.2 Secondary Operations

The Boolean composition of constraints is made complete, in the sense that any Boolean expression can be captured, by the inclusion of negation, as was done in [160], through

$$\neg h(x) \geq 0 \iff -h(x) \geq 0, \tag{6.9}$$

where one has to accept the ambiguity associated with the boundary $h(x) = 0$. This allows us to combine together constraints reflecting arbitrary Boolean expressions[1] through the application of De Morgan's Laws [132], in that $h_1 \vee h_2 = \neg(\neg h_1 \wedge \neg h_2)$.

For example, given the sets S_1, S_2, S_3 (with associated CBFs h_1, h_2, h_3, and where we use S as shorthand for the proposition $x \in S$), the CBF associated with the formula $S_1 \wedge (S_2 \vee \neg S_3)$ thus becomes

$$h_{1\wedge(2\vee\neg 3)}(x) = \min\left\{h_1(x), \max\{h_2(x), -h_3(x)\}\right\}. \tag{6.10}$$

---

[1] The one dissonant note here is given by the fact that *max* and *min* are actually non-smooth operators. As such, turning $h_{1\wedge 2}$ and $h_{1\vee 2}$ into CBFs requires some additional machinery as the CBFs from the previous chapters were all assumed to be continuously differentiable. This issue will have to be resolved in subsequent sections.

Similarly, secondary operations, like the material implication, $S_1 \Rightarrow S_2$, which is logically equivalent to $\neg S_1 \vee S_2$, and the exclusive or, $S_1 \oplus S_2$ (equivalent to $(S_1 \vee S_2) \wedge \neg(S_1 \wedge S_2)$), become

$$h_{1\Rightarrow 2}(x) = \max\left\{-h_1(x), h_2(x)\right\}, \tag{6.11}$$

$$h_{1\oplus 2}(x) = \min\left\{\max\{h_1(x), h_2(x)\}, -\min\{h_1(x), h_2(x)\}\right\}. \tag{6.12}$$

In particular, the material implication provides a way to manage situations where constraints may be violated due to a lack of feasibility. For example, the constraint $\neg S_1 \Rightarrow S_2$ encodes the proposition that the state of the system should satisfy the constraint $x \in S_1$. But, if it does not, then $x \in S_2$ should be satisfied as a fall-back.

The ability to express arbitrary Boolean compositions of constraints is vital for the formulation of a constraint-based design strategy for long-duration autonomy. To appreciate how important the ability to combine together multiple constraints really is, as well as foreshadow results in Chapter 7, recall the discussion in Chapter 3 (Equations 3.16 and 3.20) that concluded that for a robot to survive, it should satisfy the constraint $x \in \mathcal{G}_{survive}$, where this set was defined through the intersection,

$$\mathcal{G}_{survive} = \mathcal{G}_{resources} \cap \mathcal{G}_{interact} \cap \mathcal{G}_{homeostasis}. \tag{6.13}$$

The interaction constraint, as captured by the set, $\mathcal{G}_{interact}$, was afforded the additional structure,

$$\mathcal{G}_{interact} = \mathcal{G}_{separation} \cap \mathcal{G}_{connect} \cap \mathcal{G}_{topology}. \tag{6.14}$$

Equating intersections with conjunctions, these set-membership constraints can thus be replaced with the Boolean composition,

$$x \in \mathcal{G}_{resources} \ \wedge \ x \in \mathcal{G}_{interact} \ \wedge \ x \in \mathcal{G}_{homeostasis}, \tag{6.15}$$

or

$$x \in \mathcal{G}_{resources} \ \wedge \ \left(x \in \mathcal{G}_{separation} \ \wedge \ x \in \mathcal{G}_{connect} \ \wedge \ x \in \mathcal{G}_{topology}\right)$$
$$\wedge \ x \in \mathcal{G}_{homeostasis}. \tag{6.16}$$

Assuming that these sets can all be defined as super-level sets to some scalar functions, the corresponding inequality constraint becomes $h_{survive}(x) \geq 0$, where

$$h_{survive} = \min\{h_{resources}, \min\{h_{separation}, h_{connect}, h_{topology}\},$$
$$h_{homeostasis}\}, \qquad\qquad (6.17)$$

where we, for ease-of-notation, have suppressed the x-dependence among the functions. Furthermore, as min is an associative operator, the final survival constraint "simply" becomes

$$h_{survive} = \min\{h_{resources}, h_{separation}, h_{connect}, h_{topology}, h_{homeostasis}\}.$$
$$(6.18)$$

But, as we will see shortly, *just because we can write down a constraint, it does not follow that we can enforce it.* Nor even directly make technical sense of it.

6.2   Non-Smooth Barrier Functions

Now that we have established that "all" we need to do to capture full-blown Boolean composition of constraints is to introduce min and max operators, what prevents us from just deploying the framework from the previous two chapters? The clue to the obstruction resides with the way that these two functions behave whenever their constituent functions cross over each other. In the previous discussions about CBFs, we always insisted on $h \in C^1$, i.e., on the CBF being continuously differentiable. But, as seen in Figure 6.2, neither min nor max are C^1. In fact, whenever either of those operators transition from one argument function to the other, a non-differentiability is (typically) introduced. To this end, we first need to expand the CBF toolbox to encompass Non-Smooth Barrier Functions (NBFs) as well.

Though not considered in this particular context, non-smooth Lyapunov functions have been extensively studied, e.g., [30, 31, 322, 376]. Luckily, the tools developed for analyzing and characterizing such non-smooth Lyapunov functions translate quite readily and gracefully over to the NBF setting. In this chapter, we show how to extend the previously established concepts for smooth CBFs to the richer class of NBFs.

To make clear the need to modify and generalize what was previously done in the smooth case, consider, initially, the situation where there is no

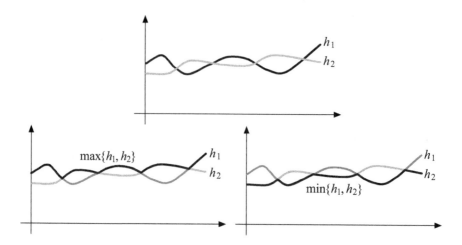

Figure 6.2: Two functions, h_1 and h_2 (Top), are plotted together with their corresponding min and max functions (Bottom). As can be noted, whenever $h_1 = h_2$, the min and max functions are typically no longer differentiable, despite h_1 and h_2 themselves being smooth.

control input, and the system is given by

$$\dot{x} = f(x), \tag{6.19}$$

where we assume that f is locally Lipschitz continuous over its entire domain. The forward invariance constraint from Chapter 4 (Equation 4.15), associated with ensuring that a continuously differentiable function, h, remains positive for all times, stated that

$$\frac{d}{dt} h(x(t)) \geq -\alpha(h(x(t))), \tag{6.20}$$

where α is a locally Lipschitz continuous, extended class $\mathcal{K}_\infty$-function.

Let us unpack the time-derivate of h a bit further, as we move towards a non-smooth version of the forward invariance constraint. Thanks to the Chain Rule, we have that

$$\frac{d}{dt} h(x) = \frac{\partial h(x)}{\partial x}^T \dot{x} = \frac{\partial h(x)}{\partial x}^T f(x) = \langle \nabla h(x), f(x) \rangle, \tag{6.21}$$

where $\langle \cdot, \cdot \rangle$ denotes inner product. And, this inner product seems like a promising place to start the non-smooth investigation. What happens when

∇h is no longer well-defined at some points, as is the case with the min and max functions? What we need to do is reinterpret what the "gradient," ∇h, actually entails.

6.2.1 Generalized Gradients

Gradients are defined through limits. For example, for scalar functions, $\phi : \mathbb{R} \to \mathbb{R}$, we have that

$$\phi'(t) = \lim_{\epsilon \to 0} \frac{\phi(t+\epsilon) - \phi(t)}{\epsilon}. \tag{6.22}$$

This construction is completely unproblematic as long as the limit is well-defined. But, what if, for example, $\phi(t) = |t|$? In that case, $\phi'(t) = 1$ if $t > 0$, and $\phi'(t) = -1$ if $t < 0$. The problem appears when $t = 0$. The limit in Equation 6.22 is different at $t = 0$ depending on whether $\epsilon \to 0$ happens from above or below, i.e., for $\phi(t) = |t|$, we have

$$\lim_{\epsilon \searrow 0} \frac{\phi(\epsilon) - \phi(0)}{\epsilon} = 1, \quad \lim_{\epsilon \nearrow 0} \frac{\phi(\epsilon) - \phi(0)}{\epsilon} = -1. \tag{6.23}$$

Which one is the "correct" gradient? The answer is both. And perhaps even everything in between, i.e., the entire interval $[-1, 1]$. In other words, what used to be a point has now become a set.

The next order of business is to generalize this observation about $\phi(t) = |t|$, which by the way can be written as the max function,

$$\phi(t) = |t| = \max\{-t, t\}, \tag{6.24}$$

to other non-smooth functions that could, for example, also originate from the repeated application of max- and min-operations to smooth functions. To this end, let $\xi : \mathbb{R}^n \to \mathbb{R}^m$ be locally Lipschitz continuous on a domain containing a zero-measure set (under the Lebesgue measure), Ω, where ξ is non-differentiable. The gradient, $\nabla \xi(x)$, is, of course, unproblematic whenever $x \notin \Omega$. But, when $x \in \Omega$, we have to revisit the idea of defining the proper limits in order to arrive at a set, rather than a single value for the gradient.

Recall that we decided to let the gradient of $\phi(t) = |t|$ (with t being scalar), at $t = 0$, be given by the entire interval $[-1, 1]$, which can be obtained by taking the convex hull of the results obtained when using two different limits, -1 and $+1$. In fact, in Equation 6.23, we actually (indirectly) specified that the two limits had to be constructed from limiting sequences obtained by letting ϵ go to zero from above and from below. An alternative way of arriving at the same result would be to not restrict oneself to only two such sequences, but instead take the convex hull of the limits of $\nabla\phi$ obtained across *all possible sequences* converging to $t = 0$, but always staying away from that particular, problematic point. This alternative characterization can be described using the following notation,

$$\text{co}\left\{\lim_{i\to\infty} \nabla\phi(t_i) \mid t_i \to 0, \ t_i \neq 0\right\} = [-1, 1]. \tag{6.25}$$

We use $\tilde{\nabla}\phi$ to denote this *generalized gradient*, and note that this idea is immediately generalizable to the function ξ, as per the formalism developed by F. H. Clarke in [85].

> **Definition 6.1** Let ξ be locally Lipschitz continuous, and let Ω be the zero-measure set where ξ is non-differentiable. Then the generalized gradient of ξ is given by
>
> $$\tilde{\nabla}\xi(x) = \text{co}\left\{\lim_{i\to\infty} \nabla\xi(x_i) \mid x_i \to x, \ x_i \notin \Omega\right\}. \tag{6.26}$$

This limiting construction for obtaining the generalized gradient is pictorially illustrated in Figure 6.3 (Left), while Figure 6.3 (Right) shows more explicitly what this looks like for the function $\min\{h_1, h_2\}$, where the generalized gradient is the set containing the singleton, "normal" gradient at all points, except those where $h_1(x) = h_2(x)$, in which case the gradient becomes fully (non-trivially) set-valued. In fact, what we end up with for min (and, analogously for max) functions is

$$\tilde{\nabla}\Big[\min\{h_1(x), h_2(x)\}\Big] = \begin{cases} \{\nabla h_1(x)\} & \text{if } h_1(x) < h_2(x) \\ \{\nabla h_2(x)\} & \text{if } h_2(x) < h_1(x) \\ \text{co}\{\nabla h_1(x), \nabla h_2(x)\} & \text{if } h_1(x) = h_2(x). \end{cases} \tag{6.27}$$

6.2.2 Set-Valued Lie Derivatives

When h is C^1, we have that $\dot{h} = \langle \nabla h, f \rangle$, with $\dot{x} = f(x)$, which in turn can be written as the Lie derivative, $L_f h(x)$. But what happens when we no longer have ∇h but $\tilde{\nabla} h$? In that case, the inner product somehow becomes an inner product between a set and f, which would potentially mean that $\dot{h}$ takes on

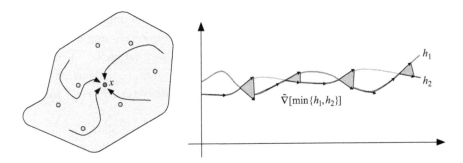

Figure 6.3: Left: A function is defined over a domain containing a zero-measure set, Ω, of points of non-differentiability (depicted as small circles). The generalized gradient is obtained by taking the convex hull of the limits obtained across the gradients of all sequences converging to x, but staying away from the problematic set, Ω. Right: The generalized gradient associated with $\min\{h_1, h_2\}$. As can be seen, when the constituent functions are different, the generalized gradient is (the set consisting of) the standard gradient, but whenever the functions are equal, it becomes a set defined by the convex hull over ∇h_1 and ∇h_2, as depicted by the dotted lines connecting the two gradients at the points of non-differentiability.

values in a set as well. To untangle what this entails, we have to take a brief detour into the world of differential inclusions, e.g., [141].

Given a set-valued map, $F : \mathbb{R}^n \to 2^{\mathbb{R}^n}$, one can let this map generate a dynamical system through the differential inclusion

$$\dot{x}(t) \in F(x(t)). \tag{6.28}$$

There is quite a bit of technical machinery required to manage such an expression. Rather than disrupting the flow of the narrative, we relegate the explanations of the various terms to footnotes over the next few paragraphs. To make sense of Equation 6.28, we first of all need to ensure that solutions exist. Without that fact firmly established, the subsequent discussion runs the risk of becoming slightly meaningless.

To get at existence of solutions, assume that F is locally bounded,[2] upper semi-continuous,[3] and $F(x)$ is non-empty, compact, and convex for each x.

---

[2] F being locally bounded means that for any x, there exists an M such that, for all x' in an open neighborhood around x, $\|\zeta\| < M$ for all $\zeta \in F(x')$. In other words, nothing too dramatic happens to the possible values that F can take on around x.

[3] For functions mapping from $\mathbb{R}$ to $\mathbb{R}$, upper semi-continuity means that for every point, x, the difference between the function values at x and $x + \epsilon$, for *positive* and small ϵ,

These properties do indeed ensure the existence (but not uniqueness) of solutions to Equation 6.28, e.g., [92]. Additionally, we need to figure out what it means for something to be a "solution" to a differential inclusion. To this end, a Carathéodory solution to the differential inclusion in Equation 6.28 is an absolutely continuous[4] trajectory, $x(t)$, such that $\dot{x}(t) \in F(x(t))$ almost everywhere.[5]

Following [92], we need to make sense of the idea of allowing for set-valued Lie derivatives by combining generalized gradients with differential inclusions. In [30], these constructs were used to analyze non-smooth Lyapunov functions, and the same tool—the set-valued Lie derivative[6]—may be applied to non-smooth barrier functions as well. The following result is taken from [92, 161], with a slightly modified notation to better fit the current context.

Lemma 6.1 Let $x \in D \subset \mathbb{R}^n$ be the solution to $\dot{x} = f(x)$ (as per Equation 6.19), and let $h : D \subset \mathbb{R}^n \to \mathbb{R}$ be locally Lipschitz continuous. Then, $h(x(t))$ is a Carathéodory solution to

$$\dot{h}(x(t)) \in \tilde{L}_f h(x(t)), \tag{6.29}$$

where the set-valued Lie derivative is given by

$$\tilde{L}_f h(x) = \{\ell \in \mathbb{R} \mid \exists v \in \tilde{\nabla} h(x) \; s.t. \; \langle v, f \rangle = \ell\}. \tag{6.30}$$

---

is "small." The generalization to set-valued mappings is slightly more involved, but the basic idea is the same. At $x \in \mathbb{R}^n$, and for all $\epsilon > 0$, there exists a $\delta > 0$, such that $F(x') \subseteq F(x) + B_\epsilon(x)$, for all $y \in B_\delta(x)$, where $B_\epsilon(x)$ and $B_\delta(x)$ are the open balls of radius ϵ and δ, respectively, centered at x. So, small changes in x produce "small" changes in the set-valued map. The one technical detail to pay attention to here is the order in which the quantifiers are applied to $B_\epsilon(x)$ and $B_\delta(x)$—this is what causes the function to be upper semi-continuous as opposed to "merely" continuous.

[4] Absolute continuity is slightly stronger than plain continuity (but weaker than Lipschitz continuity) in that an absolutely continuous function can be described by an integral over a Lebesgue integrable function.

[5] For all but a zero-measure set of times.

[6] If $\dot{x} \in F(x)$, i.e., the system dynamics is a differential inclusion, the distinction is made between strong and weak set-valued Lie derivatives. This distinction is not entirely meaningful when $\dot{x} = f(x)$, nor for the technical developments in this chapter. More details about this topic can be found, for example, in [92].

The interpretation here is that the non-smooth part of an NBF translates to $\dot{h}$ taking on values in a set defined by the set-valued Lie derivative. What this means for forward invariance is, however, not yet entirely clear.

In the smooth case, forward invariance is ensured if there exists a locally Lipschitz continuous, extended class $\mathscr{K}_\infty$-function, α, such that

$$\dot{h} \geq -\alpha(h). \tag{6.31}$$

But now, in the non-smooth setting, we have just discovered that $\dot{h}$ takes on values in a set, defined by the set-valued Lie derivative, as per Lemma 6.1. A natural idea is to ensure that no matter what element in that set is "chosen," $\dot{h}$ is still greater than some $-\alpha$. This can be achieved by making sure that $\min \tilde{L}_f h(x) \geq -\alpha(h(x))$, where the minimum is taken over the set defined by the set-valued Lie derivative in Equation 6.30. This concept is illustrated in Figure 6.4. And, it turns out that minimizing over this set is exactly what is needed to ensure forward invariance in the non-smooth case [160].

Theorem 6.1 Let $h : D \subset \mathbb{R}^n \to \mathbb{R}$ be a locally Lipschitz continuous function whose super-level set, $\mathscr{C} = \{x \in D \mid h(x) \geq 0\}$, is non-empty. If there exists a locally Lipschitz continuous, extended class $\mathscr{K}_\infty$-function, α, such that the minimum over the set-valued Lie derivative satisfies

$$\min \tilde{L}_f h(x) \geq -\alpha(h(x)), \ \forall x \in D, \tag{6.32}$$

then h is a valid NBF in that $h(x(t)) \geq 0$ is ensured, i.e., $h \geq 0$ is forward invariant.

The key upshot from Theorem 6.1 is that the high-level principle of ensuring that $\dot{h}$ does not become too negative as the function gets closer to $h = 0$ still holds. The difference is that this has to hold for *all* values $\dot{h}$ may take on at a given point, through its associated differential inclusion. This, in turn, is ensured by minimizing over an entire set of possible values, i.e., $\dot{h}$ *cannot be too negative close to the boundary of the safe set.*

6.3 Min/Max Barrier Functions

Now that we have the machinery in place to express set-valued Lie derivatives and their relevance to non-smooth barrier functions, we return to what

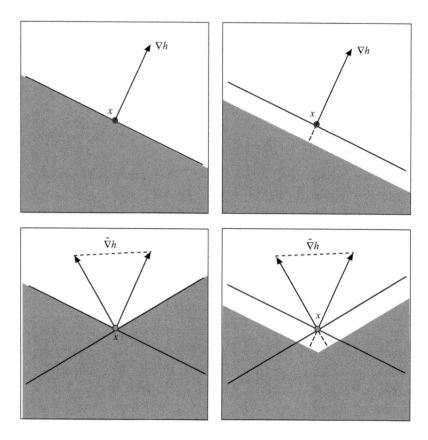

Figure 6.4: Top Left: The non-shaded region shows the set where f may take on values in order to render the inner product $\langle \nabla h, f \rangle \geq 0$ in the smooth case, i.e., when $h \in C^1$. Top Right: The same situation is depicted with the twist that the shaded region is offset by a fixed amount, which enlarges the region where f may reside in order to ensure that $\langle \nabla h, f \rangle \geq -\alpha(h)$, where $\alpha(h) \geq 0$ captures the offset. Bottom Left: Here h is no longer differentiable and the generalized gradient is given by the convex hull of two vectors. As the inner product needs to be non-negative for all elements in the convex hull, the non-shaded region, where f can take on values such that $\min\{\langle \tilde{\nabla} h, f \rangle\} \geq 0$, has shrunk somewhat. Bottom Right: The introduction of an offset, such that $\min\{\langle \tilde{\nabla} h, f \rangle\} \geq -\alpha(h)$, enlarges the non-shaded region as compared to the non-offset situation.

precipitated this technical detour, namely Boolean composition of barrier functions in general, and max and min barrier functions in particular.

For min and max functions (operating on smooth argument functions), the points of possible non-differentiability occur when there are multiple, constituent functions taking on the same value. For bookkeeping reasons, it proves handy to call out those functions explicitly. To that end, assume that we are looking at combining (through min or max operators) a finite set of functions $h_i : D \subset \mathbb{R}^n \to \mathbb{R}$, $i \in [k]$, where we, as before, let $[k]$ denote the index set $\{1, \ldots, k\}$.

As already established, max represents disjunction, i.e., the Boolean $\vee$ operation. Let $h_{[k]}^{\max} : D \subset \mathbb{R}^n \to \mathbb{R}$ be this max function,

$$h_{[k]}^{\max}(x) = \max_{i \in [k]} \left\{ h_i(x) \right\}. \tag{6.33}$$

Obviously, if we evaluate this along trajectories, then, for each t, there exists at least one $j \in [k]$ such that $h_j(x(t)) = h_{[k]}^{\max}(x(t))$. And, if $h_{[k]}^{\max}(x(t))$ is a *valid* NBF, then we furthermore must have that $h_j(x(t)) \geq 0$ at that particular time, t. But, as already noted, there are points where multiple functions take on the same value, and we let the corresponding index set be given by

$$\mathscr{I}^{\max}(x) = \left\{ j \in [k] \mid h_j(x) = h_{[k]}^{\max}(x) \right\}. \tag{6.34}$$

Analogously, since min represents conjunction, i.e., the Boolean $\wedge$ operation, we let

$$h_{[k]}^{\min}(x) = \min_{i \in [k]} \left\{ h_i(x) \right\} \quad \text{and} \quad \mathscr{I}^{\min}(x) = \left\{ j \in [k] \mid h_j(x) = h_{[k]}^{\min}(x) \right\}. \tag{6.35}$$

6.3.1 Boolean Composition of Barrier Functions

We now dig in a bit deeper into the question of how to map these Boolean compositional NBFs to the set-valued results from Section 6.2. In particular, the statement in Theorem 6.1 involves taking the min over a convex hull in order to ensure that the required inequality constraint is satisfied.

Luckily, we do not actually have to spend too much computational effort on this minimization endeavor as it suffices, as shown in [160], to only consider the gradients (continuous and well-defined) of the constituent functions themselves. This is due to the fact that the minimization of a linear cost, constrained to a convex hull, is achieved at one of the extreme points in the

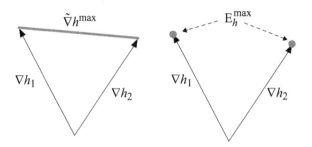

Figure 6.5: Rather than minimizing over the entire set, it suffices to consider the extreme points, as per Theorem 6.2., i.e., $\min \tilde{\nabla} h^{\max} = \min \mathrm{co}\,\{\nabla h_1, \nabla h_2\} = \min \mathscr{E}_h^{\max} = \min\{\nabla h_1, \nabla h_2\}$.

set, e.g., [58, 358], as illustrated in Figure 6.5. We can thus draw upon the following, computational result, found in [160].

Theorem 6.2 Let $h_i : D \subset \mathbb{R}^n \to \mathbb{R}$, $i \in [k]$, be a finite set of locally Lipschitz continuous functions, and let $h_{[k]}^{\max}$ and $\mathscr{I}^{\max}$ be defined as in Equations 6.33 and 6.34, respectively. For each $x \in D$, consider the set-valued map, $\mathscr{E}_h^{\max} : D \subset \mathbb{R}^n \to 2^{\mathbb{R}^n}$, given by

$$\mathscr{E}_h^{\max}(x) = \bigcup_{j \in \mathscr{I}^{\max}(x)} \{\nabla h_j(x)]\}. \tag{6.36}$$

If there exists a locally Lipschitz continuous, extended class $\mathscr{K}_\infty$-function, α, such that for every $x \in D$ and $\xi \in \mathscr{E}_h^{\max}(x)$, $\langle \xi, f(x) \rangle \geq -\alpha(h_{[k]}^{\max}(x))$, then $h_{[k]}^{\max}$ is a valid NBF to the system, $\dot{x} = f(x)$.

Theorem 6.2 extends, not surprisingly, in a completely analogous fashion to the min-situation as well (using $h^{\min}$, $\mathscr{I}^{\min}$, $\mathscr{E}^{\min}$), and it gives us a generally efficient computational tool for ensuring forward invariance when the barrier function is made up of repeated applications of min and max operations.[7] This, in turn, allows us to handle arbitrary Boolean compositions of barrier functions.

---

[7]Even though only the constituent functions need to be considered, the depth of the Boolean expression has the potential to be computationally cumbersome nonetheless. If all min/max functions are taken over pairs of functions, and the depth of the expression is d,

The one caveat to pay attention to is that so far we have only covered the situation where $\dot{x} = f(x)$, i.e., for systems that evolve autonomously, without any external control inputs. The next (and final) ingredient in the Boolean composition recipe is thus the addition of a control term, and to consider control-affine rather than autonomous dynamical systems.

As was done in Chapter 4, let the robot dynamics be given by the control-affine system

$$\dot{x} = f(x) + g(x)u, \tag{6.37}$$

where u is the control input that must be selected so as to advance the overall mission, subject to some safety constraint, such as the robot ecological survival constraint "Don't die!" With the Boolean composition framework largely in place, we finally have some hope of being able to breathe life and meaning into this, until now, somewhat nebulous survival constraint.

Assume that f and g in Equation 6.37 are locally Lipschitz continuous, and that we need to satisfy $h_1 \geq 0$ *or* $h_2 \geq 0$, i.e., we need $h_{[2]}^{\max} \geq 0$. If $h_1(x) > h_2(x)$, we have $h_{[2]}^{\max} = h_1$, and we simply require u to satisfy is the old CBF constraint,

$$\langle \nabla h_{[2]}^{\max}(x), f(x) + g(x)u \rangle \geq -\alpha(h_{[2]}^{\max}(x)), \tag{6.38}$$

where the gradient, $\nabla h_{[2]}^{\max}$, is of the normal rather than the generalized variety.

If $h_1(x) = h_2(x)$, we, per Theorem 6.2, dodge the full-fledged convex hull and instead consider the smaller set (as illustrated in Figure 6.5),

$$\mathcal{E}_h^{\max}(x) = \{\nabla h_1(x)\} \cup \{\nabla h_2(x)\}. \tag{6.39}$$

Subsequently, the task becomes that of picking a control input, u, such that

$$\langle \xi, f(x) + g(x)u \rangle \geq -\alpha(h_{[2]}^{\max}(x)), \tag{6.40}$$

for all $\xi \in \mathcal{E}_h^{\max}$.

The only thing remaining is to put all of this together to be able to address complex Boolean expressions, as was done automatically and recursively in [161]. However, rather than unleashing the somewhat involved machinery for general such expressions, let us instead elucidate what this would look like for two of the secondary expressions in Section 6.1, namely $h_{1 \wedge (2 \vee \neg 3)}$ and

---

then, in the worst case, the total number of values that must be considered is 2^d, which is, potentially, a large number. In [161], this number is significantly reduced by a clever down-select of the values needed to be taken into account.

$h_{1\oplus2}$, and take these two examples as evidence that we can address any other Boolean expression coming our way.

$h_{1\wedge(2\vee\neg3)}$

As per Equation 6.10, we have that

$$h_{1\wedge(2\vee\neg3)}(x) = \min\Big\{h_1(x), \max\{h_2(x), -h_3(x)\}\Big\}, \tag{6.41}$$

which spawns a few different cases, where the salient differences are called out in boldface.

Case 1: $h_1 < \max\{h_2, -h_3\}$
     CBF Constraint: $\langle\nabla\mathbf{h_1}, f+gu\rangle \geq -\alpha(h_{1\wedge(2\vee\neg3)})$
Case 2: $h_1 > \max\{h_2, -h_3\}$
     Case 2a: $h_2 > -h_3$
         CBF Constraint: $\langle\nabla\mathbf{h_2}, f+gu\rangle \geq -\alpha(h_{1\wedge(2\vee\neg3)})$
     Case 2b: $h_2 < -h_3$
         CBF Constraint: $\langle-\nabla\mathbf{h_3}, f+gu\rangle \geq -\alpha(h_{1\wedge(2\vee\neg3)})$
     Case 2c: $h_2 = -h_3$
         CBF Constraint: $\langle\xi, f+gu\rangle \geq -\alpha(h_{1\wedge(2\vee\neg3)}), \ \forall\xi \in \{\nabla\mathbf{h_2}, -\nabla\mathbf{h_3}\}$
Case 3: $h_1 = \max\{h_2, -h_3\}$
     Case 3a: $h_2 > -h_3$
         CBF Constraint: $\langle\xi, f+gu\rangle \geq -\alpha(h_{1\wedge(2\vee\neg3)}), \ \forall\xi \in \{\nabla\mathbf{h_1}, \nabla\mathbf{h_2}\}$
     Case 3b: $h_2 < -h_3$
         CBF Constraint: $\langle\xi, f+gu\rangle \geq -\alpha(h_{1\wedge(2\vee\neg3)}), \ \forall\xi \in \{\nabla\mathbf{h_1}, -\nabla\mathbf{h_3}\}$
     Case 3c: $h_2 = -h_3$
         CBF Constraint: $\langle\xi, f+gu\rangle \geq -\alpha(h_{1\wedge(2\vee\neg3)}),$
                 $\forall\xi \in \{\nabla\mathbf{h_1}, \nabla\mathbf{h_2}, -\nabla\mathbf{h_3}\}$

$h_{1\oplus2}$

Equation 6.12 states that

$$h_{1\oplus2}(x) = \min\Big\{\max\{h_1(x), h_2(x)\}, -\min\{h_1(x), h_2(x)\}\Big\}, \tag{6.42}$$

with the corresponding cases being

Case 1: $h_1 > h_2$

  Case 1a: $h_1 < -h_2$

    CBF Constraint: $\langle \nabla \mathbf{h_1}, f + gu \rangle \geq -\alpha(h_{1\oplus 2})$

  Case 1b: $h_1 > -h_2$

    CBF Constraint: $\langle -\nabla \mathbf{h_2}, f + gu \rangle \geq -\alpha(h_{1\oplus 2})$

  Case 1c: $h_1 = -h_2$

    CBF Constraint: $\langle \xi, f + gu \rangle \geq -\alpha(h_{1\oplus 2}), \ \forall \xi \in \{\nabla \mathbf{h_1}, -\nabla \mathbf{h_2}\}$

Case 2: $h_1 < h_2$

  Case 2a: $-h_1 < h_2$

    CBF Constraint: $\langle -\nabla \mathbf{h_1}, f + gu \rangle \geq -\alpha(h_{1\oplus 2})$

  Case 2b: $-h_1 > h_2$

    CBF Constraint: $\langle \nabla \mathbf{h_2}, f + gu \rangle \geq -\alpha(h_{1\oplus 2})$

  Case 2c: $-h_1 = h_2$

    CBF Constraint: $\langle \xi, f + gu \rangle \geq -\alpha(h_{1\oplus 2}), \ \forall \xi \in \{-\nabla \mathbf{h_1}, \nabla \mathbf{h_2}\}$

Case 3: $h_1 = h_2$

  Case 3a: $h_1 < 0$

    CBF Constraint: $\langle \xi, f + gu \rangle \geq -\alpha(h_{1\oplus 2}), \ \forall \xi \in \{\nabla \mathbf{h_1}, \nabla \mathbf{h_2}\}$

  Case 3b: $h_1 > 0$

    CBF Constraint: $\langle \xi, f + gu \rangle \geq -\alpha(h_{1\oplus 2}), \ \forall \xi \in \{-\nabla \mathbf{h_1}, -\nabla \mathbf{h_2}\}$

  Case 3c: $h_1 = 0$

    CBF Constraint: $\langle \xi, f + gu \rangle \geq -\alpha(h_{1\oplus 2}),$
$$\forall \xi \in \{\nabla \mathbf{h_1}, -\nabla \mathbf{h_1}, \nabla \mathbf{h_2}, -\nabla \mathbf{h_2}\}$$

Regardless of whether the barrier function is smooth or non-smooth, we will use the established notation, $S_h[x; u] \geq 0$, even though we are now fully aware of the fact that this innocent-looking expression might hide a number of slightly more involved components, such as set-valued Lie derivates and, at times, quite a few different cases to consider computationally.

6.3.2 Navigation Example

Although the primary motivation for pursuing composed constraints comes from the long-duration autonomy context, the ability to combine together constraints using Boolean logic is quite useful also in more traditional robotics settings. For example, the navigation problem, at its core, concerns itself with how to move a robot from Point A to Point B, without colliding with obstacles in the environment, e.g., [83, 225]. The multi-robot version of this problem just involves more Points A and B, and avoiding inter-robot collisions as well, e.g., [43, 420, 423], which we consider here through the lens of composed constraints.

Consider a team of N mobile robots, with planar positions, $x_i \in \mathbb{R}^2$, $i \in [N]$, with the combined state $x = [x_1^T, \ldots, x_N^T]^T \in \mathbb{R}^{2N}$. Embracing the navigation theme, assume that each robot has been tasked with the primary mission of driving to a particular goal location, $x_{goal} \in \mathbb{R}^{2N}$, under the nominal control input,

$$u_{nom,i} = \kappa (x_{goal,i} - x_i), \quad i \in [N], \tag{6.43}$$

for some positive P-gain, $\kappa > 0$. Here we have assumed that the robot velocities are directly controlled, i.e., $\dot{x}_i = u_i$, or, viewed as an ensemble, $\dot{x} = u$, where $u = [u_1^T, \ldots, u_N^T]^T \in \mathbb{R}^{2N}$.

To avoid collisions with other robots and adhering to the Reynolds' Boids terminology, the compositional NBF applies to each pair of robots

$$h_{separation}(x) = \bigwedge_{i,j\in[N],i\neq j} \left[\|x_i - x_j\|^2 - D_s^2 \right], \tag{6.44}$$

where the large $\wedge$-symbol is shorthand for sequential conjunctions, and where $D_s > 0$ is a given, inter-robot safety distance.[8] Similarly, each robot should also avoid collisions with p circular obstacles, located at $o_j \in \mathbb{R}^2$, $j \in [p]$, via the NBF

$$h_{obstacle}(x) = \bigwedge_{i=1}^{N} \bigwedge_{j=1}^{p} \left[\|x_i - o_j\|^2 - D_o^2 \right], \tag{6.45}$$

where $D_o > 0$. The resulting, navigational barrier function is thus given by

$$h_{navigation}(x) = h_{separation}(x) \wedge h_{obstacle}(x). \tag{6.46}$$

---

[8]For notational compactness, we have additionally embraced the convention that $h_1(x) \wedge h_2(x)$ is shorthand for the conjunctive proposition $h_1(x) \geq 0 \wedge h_2(x) \geq 0$, and similarly for the disjunction and negation operators.

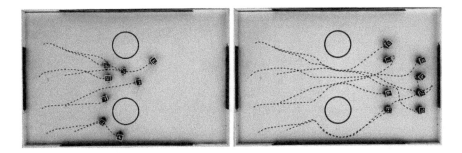

Figure 6.6: A group of 8 robots successfully navigate around a pair of obstacles (circles) to their goal locations, while avoiding inter-robot collisions, as per the navigation constraint, $h_{navigation} \geq 0$ in Equation 6.46. Source: Fig. 1 in Glotfelter et al. (2017).

Following the optimization-based framework laid out in Chapter 4, the actual control signal should be as close as possible to u_{nom}, and the corresponding quadratic programming problem to be solved, at each moment in time, becomes

$$\min_{u} \|u - u_{nom}\|^2, \tag{6.47}$$

subject to the constraint

$$S_{h_{navigation}}[x; u] \geq 0. \tag{6.48}$$

Figure 6.6 displays what happens when 8 mobile robots solve this constrained optimization problem, with $p = 2$, i.e., over an area populated with two obstacles. Additionally, Figure 6.7 verifies that the NBF in Equation 6.46 indeed remains positive for the duration of the experiment, as prescribed, i.e., all component barrier functions are simultaneously satisfied. Furthermore, as a result of the minimally invasive surgery done to u_{nom}, thanks to the least-square cost, $\|u - u_{nom}\|^2$, the robots also arrive successfully at their goal destinations.

6.4 Connectivity-Preserving Coordinated Control

In Chapter 4, Section 4.3, a more involved collision-avoidance scenario was considered, where the robots were no longer assumed to be able to perform the physical impossibility of instantaneously changing their velocities. What this means is that maximum acceleration limit has to be taken into account when

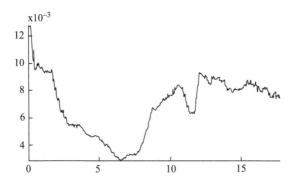

Figure 6.7: The value of the NBF in Equation 6.46, $h_{navigation}$, is plotted as a function of time. Since the NBF is positive for all times, the different safety objectives are simultaneously satisfied, as expected [423]. Source: Fig. 2 in Glotfelter et al. (2017).

designing the barrier functions. We here revisit this setup in the context of Boolean composition of arbitrary constraints, in order to give the developed framework a more thorough vetting. Additionally, we not only require that the robots avoid collisions, but also that the underlying information-exchange network stays connected, as per the discussion about team-level connectivity constraints in Chapter 3.

6.4.1   Composite Safety and Connectivity Barrier Functions

Following the program from Section 4.3, let the multi-robot system, once again, be given by N planar, mobile robots, with dynamics

$$\begin{bmatrix} \dot{p}_i \\ \dot{v}_i \end{bmatrix} = \begin{bmatrix} 0 & I_{2\times 2} \\ 0 & 0 \end{bmatrix} \begin{bmatrix} p_i \\ v_i \end{bmatrix} + \begin{bmatrix} 0 \\ I_{2\times 2} \end{bmatrix} u_i, \quad i \in [N], \tag{6.49}$$

where the state is now four-dimensional, $x_i = [p_i^T, v_i^T]^T \in \mathbb{R}^4$, with $p_i \in \mathbb{R}^2$, $v_i \in \mathbb{R}^2$, and $u_i \in \mathbb{R}^2$ corresponding to the position, velocity, and input (acceleration command) of Robot i, $i \in [N]$. Additionally, to remain physically faithful, the acceleration is assumed to be bounded by $\|u_i\| \leq a$.

In Section 4.3, it was established that the pairwise, collision-free safe set, S_{ij}, was given by

$$S_{ij} = \left\{ (x_i, x_j) \in \mathbb{R}^8 \mid h_{separation}^{pair}(x_i, x_j) \geq 0 \right\}, \forall i \neq j, \tag{6.50}$$

where the CBF defining this super-level set took on the form

$$h^{pair}_{separation}(x_i, x_j) = 2\sqrt{a(\|\Delta p_{ij}\| - D_s)} + \frac{\Delta p_{ij}^T}{\|\Delta p_{ij}\|} \Delta v_{ij}. \qquad (6.51)$$

Here we have explicitly called out the fact that the constraint is a pairwise constraint, through the use of the superscript. Additionally, $\Delta p_{ij} = p_i - p_j \in \mathbb{R}^2$, $\Delta v_{ij} = v_i - v_j \in \mathbb{R}^2$, and D_s is the required inter-robot safety distance. And, just as in Equation 6.44 in the previous section, the combined safety barrier function becomes

$$h_{separation}(x) = \bigwedge_{i,j \in [N], i \neq j} h^{pair}_{separation}(x_i, x_j). \qquad (6.52)$$

This means that the collision-avoidance component looks much the same as in the previous section, despite the fact that the underlying barrier functions are significantly more involved.

The question of network connectivity requires a bit more thought, as compared to collision-avoidance, as connectivity is not just a local, pairwise property, but a global, team-level property, e.g., [276]. It is, however, possible to simplify the connectivity problem to a set of pairwise constraints if we assume that the multi-robot network is supposed to maintain a particular, static, and connected network as a spanning subgraph of the overall network graph.

Let $\mathcal{G} = (V, E)$ be the prespecified, required connectivity graph, where $V = \{1, 2, \ldots, N\}$ is the set of robots, and $E \subset V \times V$ is the required edge set. Moreover, we assume that the actual network is given by a *disk-graph* [263], in the sense that two robots are connected if they are within distance D_c of each other. (This disk-graph concept is illustrated in Figure 6.8.) What we need to enforce, in this particular case, is that $(i, j) \in E \Rightarrow \|p_i - p_j\| \leq D_c$, i.e., the link between Robots i and j should be maintained by having the robots stay sufficiently close together throughout the maneuver.[9]

This construction allows us to translate the global connectivity question to a set of pairwise constraints. In fact, these pairwise constraints can be directly obtained by considering the maximum acceleration needed to avoid breaking connectivity, i.e.,

$$h^{pair}_{connect}(x) = 2\sqrt{a(D_c - \|\Delta p_{ij}\|)} - \frac{\Delta p_{ij}^T}{\|\Delta p_{ij}\|} \Delta v_{ij} \geq 0, \; (i, j) \in E, \qquad (6.53)$$

---

[9] The reason why this is not an equivalence relationship is that it is perfectly fine to have $\|p_i - p_j\| \leq D_c$ even if $(i, j) \notin E$. In other words, the specified graph, $\mathcal{G}$, should be a spanning *subgraph* to the actual disk-graph network.

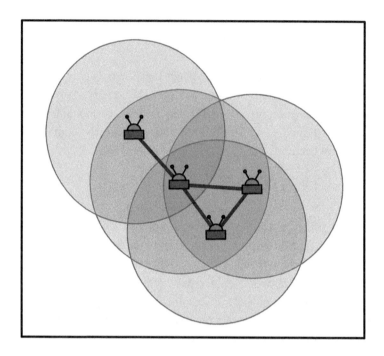

Figure 6.8: Disk-graph network. Two robots are connected if they are sufficiently close to each other, i.e., $(i, j) \in E_{disk} \Leftrightarrow \| p_i - p_j \| \leq D_c$, for some cutoff distance D_c.

whose similarity to Equation 6.52 is no coincidence as the collision-avoidance constraint, $h_{separation}^{pair} \geq 0$, ensures that the robots do not get too close, while $h_{connect}^{pair} \geq 0$ makes sure that they do not stray too far from each other.

The composite connectivity constraint becomes

$$h_{connect}(x) = \bigwedge_{(i,j) \in E} h_{connect}^{pair}(x_i, x_j). \tag{6.54}$$

Gathering everything together, the constraint needed to formally ensure both collision-free motions and connectivity maintenance is given by the following expression from [423],

$$h_{static}(x) = h_{separation}(x) \wedge h_{connect}(x), \tag{6.55}$$

where we use the subscript $static$ to explicitly call out the fact that the connectivity graph, $\mathcal{G}$, that should be maintained is static, i.e., a fixed set of edges are to be maintained at all times, as opposed to the more general connectivity constraints introduced in Chapter 3. Obviously, we need $D_s < D_c$ for solutions to exist since the robots must be farther apart from each other than D_s, yet closer than D_c.

It should be noted that significant work has been done on trying to achieve both of these objectives simultaneously, e.g., [197, 437, 439]. But, the beauty of the proposed approach, based on Boolean composition of constraints, is that the particulars of the solution are not explicitly specified in that the robots are free to do whatever they want to, as long as the constraint is satisfied.

6.4.2 Maintaining Dynamic Connectivity Graphs

If the requirement is for the network to stay connected, and the solution involves maintaining a particular, static network, it seems like we are overly restricting the possible robot motions by not allowing the network to evolve dynamically. In fact, due to changing robot positions and possibly even deployment environments it, in general, makes sense to allow the robots to switch between different connectivity graphs, as pointed out in [423].

Up to this point, the composite, overarching constraint has been given purely by conjunctions (logical ANDs), but to get to dynamic connectivity, disjunctions (ORs) are needed as any one of a number of possible graph topologies will suffice. Let $\mathcal{G} = \{\mathcal{G}_1, \mathcal{G}_2, \ldots, \mathcal{G}_M\}$ denote the set of all such allowable connectivity graphs, where $\mathcal{G}_i = (V, E_i)$, $i \in [M]$. To stay connected, the team needs to be arranged according to at least one of these allowable connectivity graphs [216].

The barrier function that encodes the dynamic connectivity graph requirement is

$$h_{connect}(x) = \bigvee_{k \in [M]} \bigwedge_{(i,j) \in E_k} h_{connect}^{pair}(x_i, x_j). \tag{6.56}$$

A four-robot instantiation of this concept is depicted in Figure 6.9. Here, Robots 2 and 3 are expected to support a network link between them, thereby acting as a *backbone* to the network. Robots 1 and 4, on the other hand, can be connected to either Robot 2 or Robot 3 (or both). The result is the set of four graphs, shown in Figure 6.9, of which at least one has to be a spanning subgraph to the actual disk-graph network.[10]

---

[10]The reason why we do not have to enumerate more graphs with the required property is that they all have at least one of the four graphs in Figure 6.9 as a spanning subgraph.

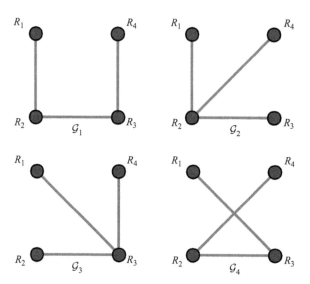

Figure 6.9: The four graphs needed to ensure that Robots 2 and 3 are always connected (thereby providing a network backbone), while Robots 1 and 4 are connected to either 2 or 3 (or both). All other graphs satisfying these requirements have at least one of the four graphs in the figure as a spanning subgraph.

Combining $h_{connect}$ in Equation 6.56 with the collision-avoidance constraint in Equation 6.52 results in the full-blown, composite constraint,

$$h_{dynamic}(x) = h_{separation}(x) \wedge h_{connect}(x). \tag{6.57}$$

To see this connectivity maintenance procedure in action, consider the situation where four robots are to follow a set of nominal paths, as shown in Figure 6.10. The nominal controller to be used in the constrained optimization problem,

$$\min_{u} \|u - u_{nom}\|^2, \tag{6.58}$$

subject to

$$S_{h_{dynamic}}[x; u] \geq 0, \tag{6.59}$$

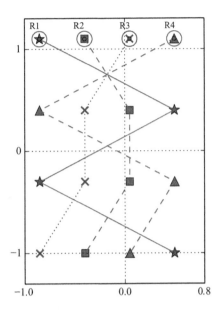

Figure 6.10: Planned waypoints for four robots. Here, R_i denotes Robot i, $i \in [4]$, while the lines represent the nominal trajectories to be executed by the robots through the use of the nominal controller. Source: Fig. 1 in Wang et al. (2016).

is given by a proportional-integral tracking controller for tracking the nominal paths in order to make the waypoints be visited sequentially. Additionally, let the set of possible connectivity graphs, $\tilde{\mathcal{G}} = \{\mathcal{G}_1, \mathcal{G}_2, \mathcal{G}_3, \mathcal{G}_4\}$, be given by the four graphs in Figure 6.9.

In the first of the experiments, only the collision-avoidance constraint, $S_{h_{separation}}[x; u] \geq 0$, is enforced and, as shown in Figure 6.11, all inter-robot distances are always larger than the safety distance, D_s, i.e., no collisions occur. Figure 6.12 displays snapshots taken by an overhead camera together with the associated robot trajectories. As can be seen, all robots successfully visit their specified waypoints without colliding. Note, however, that without the connectivity constraints, the robot team sometimes gets disconnected, e.g., the team splits up into two disconnected groups.

During the second experiment, the full-blown, composite safety and connectivity barrier functions are wrapped around the waypoint controller, i.e., the same nominal controller is used in the least-squares cost as in the first experiment, while the constraint now becomes $S_{h_{dynamic}}[x; u] \geq 0$. A snapshot of the dynamic connectivity maintenance experiment is provided in

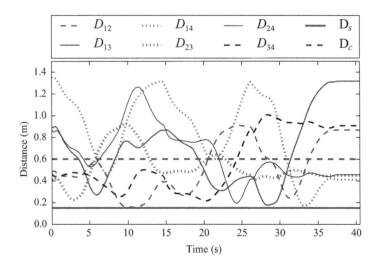

Figure 6.11: Evolution of inter-robot distances, where D_{ij} is the distance between Robots i and j. In the experiment, the two relevant distances were set to $D_s = 0.15$m and $D_c = 0.6$m, respectively. Around the 22s mark, the distance between Robots 2 and 3 becomes greater than D_c, i.e., the backbone gets disconnected in the absence of a connectivity-enforcing constraint. Source: Fig. 2 in Wang et al. (2016).

Figure 6.13, which shows that the robot team does indeed satisfy all the safety and connectivity requirements specified by the safety and connectivity barrier functions. *This remarkably complex and sophisticated behavior is obtained "simply" by specifying the appropriate constraint.* Note also that the robots visit all specified waypoints except the last one. This failure to reach the last set of waypoints is due to the fact that this last set violates the connectivity constraint, i.e., Robot 1 cannot reach its waypoint without breaking its connectivity to either Robot 2 or Robot 3. This observation highlights that not all higher-level objectives are perfectly compatible with the safety and connectivity constraints.

With that, the necessary technical machinery is complete, which also wraps up Part II of the book. We can now express safety and other pertinent features through barrier functions of varying degrees of complexity. What is left is to unleash this machinery on the connections between ecology and long-duration autonomy, conceptually established in Part I. The last three chapters in Part III will be devoted to this

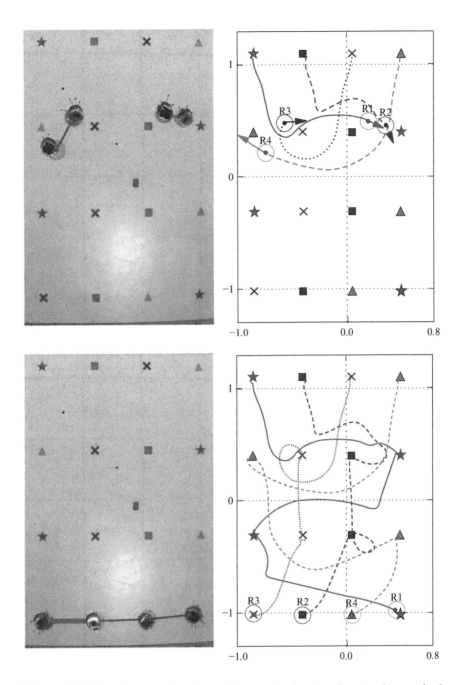

Figure 6.12: Experiment where four robots are trying to adhere to the nominal waypoint controller, subject only to the safety constraint, $h_{separation} \geq 0$. The left pictures are taken by an overhead camera. The star, square, cross, and triangular markers representing the waypoints are projected onto the ground. The right figures display the robot trajectories and their current positions. Source: Fig. 3 in Wang et al. (2016).

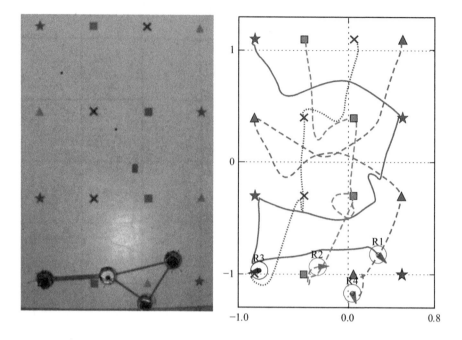

Figure 6.13: An experiment where the four robots are following the nominal waypoint controller, subject to both safety and connectivity barrier constraints. As prescribed, the robots avoid collisions, yet remain connected throughout the experiment. Source: Fig. 4 in Wang et al. (2016).

very issue. The next chapter (Chapter 7) formulates *robot ecology* as a precise, mathematical framework for controlling teams of robots. The concluding two chapters contain different instantiations of this construction. An energy-aware robot—the SlothBot—is introduced in Chapter 8 for achieving persistent environmental monitoring. Following this, recruitable and taskable robot teams in the Robotarium, under the banner of *autonomy-on-demand*, is the topic of Chapter 9.

III   Robots in the Wild

7 Robot Ecology

Now that we have made technical sense of the idea of constraint-based control design, we have the means to design controllers that allow for robots to be present and sustain themselves in natural environments over long time-scales. Captured through the constraint $x \in \mathscr{C}_{survive}$ (as vaguely formulated in Part I), or $S_{h_{survive}}[x; u] \geq 0$ (as intermittently hinted at in Part II), we are properly geared up and ready to express this constraint as a Boolean composition of multiple barrier functions. In fact, as the title of this book is *Robot Ecology: Constraint-Based Design for Long-Duration Autonomy*, we can probably declare success on the subtitle already, i.e., we have a formal framework in place for *constraint-based design for long-duration autonomy*. But is it *robot ecology*? Probably not yet.

To reach the robot ecology bar, one has to embrace the five principles arrived at in Chapter 3 (Section 3.3). For ease of reference, we here recall what these principles entail, both in terms of their ecological originators, and their robotic interpretations.

Robot Ecology Principles

P1: Maximize the Fitness Set
Interpretation: Avoid overfitting solutions to any particular set of environmental conditions;

P2: Ideal Free Distributions
Interpretation: Arrange the robots to make them recruitable and taskable to participate in a wide variety of externally specified missions;

P3: Carrying Capacity
Interpretation: The tight coupling between robot and its environment must be part of the framework;

P4: Collaborative Interactions
Interpretation: The robots should be able to work together in a safe yet productive manner; and

P5: Purposeful Expenditure of Energy
Interpretation: Survival is a prerequisite to thriving, i.e., power-management beyond a single battery charge must be an integral part of the design paradigm.

Of these five principles, the constraint-based survival formalism supports and addresses at least three directly (*P1, P3, P4*), while the addition of per-sistified tasks, coupled with minimum-energy controllers, connects readily to *P5*. But, what is missing is the final formulation of the survivability constraint itself, as well as a more principled way of encapsulating performance-based considerations, such as externally provided tasks, leading to recruitable and taskable teams, as per *P2*. This must be done while ensuring that "taskability" does not get in the way of the power-management requirement, i.e., while keeping a careful eye on the *Purposeful Expenditure of Energy* principle.[1] This chapter wraps up the robot ecology story by first developing the sur-vivability constraint(s) followed by a structured and systematic approach to formulating goals as constraints. This, in turn, allows us to keep the spec-ifications uniform across both tasks and survivability considerations, while supporting a variety of tasks.

---

[1] Note that these robotic interpretations of the ecological principles are *not* direct map-pings from ecology to robotics. Such an endeavor would constitute ecologically inspired robotics, which is not what robot ecology is ultimately about. Rather, these interpretations delineate key aspects of robot ecology, grounded (sometimes loosely and other times quite crisply) in ecology.

7.1 Constraints From Behavioral Ecology

Richness of behavior in nature is, as we have seen repeatedly, largely driven by constraints, e.g., [107]. This influence even goes beyond behaviors. Phenotype configurations, as interpreted more broadly to include an animal's physical appearance, are likewise, at least partially, constraint-driven [107, 308, 357]. As established in Part II, these connections between constraints and phenotypes can be placed on firm mathematical footing through the use of barrier functions. The resulting constraint-based control design framework can be used to generate the requisite, rich class of provably correct robot behaviors needed to support a number of long-duration autonomy tasks, such as having robots cover areas, assemble shapes, or maintain network connectivity.

7.1.1 Constituent Constraints

In this section, we gather together the constraints used as motivating examples throughout Part II, and combine them together in different ways to arrive at a formulation of the, as of yet, somewhat elusive survivability constraints outlined in Chapter 3.[2] In other words, we have already produced the puzzle pieces and we know what the final picture should look like. All that remains is to figure out how to fit the pieces together. To that end, step one is to take inventory over what pieces we have at our disposal.

In Chapter 4 (Section 4.2), a number of promising, constituent survivability constraints were cataloged. As seen, the particulars of these constraints hinge on the system dynamics, and for the sake of notational clarity and parsimoniousness, we express these constraints for robots whose dynamics are given by

$$x_i = \begin{bmatrix} p_i \\ E_i \end{bmatrix}, \quad \dot{x}_i = \begin{bmatrix} u_i \\ F(x_i) \end{bmatrix}, \quad i \in [N]. \tag{7.1}$$

Here, $p_i \in \mathbb{R}^d$ is the position of Robot i in d dimensions, $E_i \in \mathbb{R}$ is its energy level,[3] and $u_i \in \mathbb{R}^d$ the control input, as per Equations 4.23, 4.24, and 5.5.

---

[2]The reason for using the plural form is that different survivability constraints may be appropriate in different settings. As such, a one-size-fits-all survivability constraint is not a realistic, nor even desirable pursuit.

[3]In Equation 5.5, the energy dynamics are time-varying, i.e., $\dot{E}_i = F(x_i, t)$. This omission in Equation 7.1 does not change anything as we know how to handle the time-varying case as well, thanks to the developments in Chapter 5.

The generalization to control-affine systems is formally straightforward. As long as we can construct the relevant outputs,

$$p_i = \eta_p(x_i) \in \mathbb{R}^d \quad \text{and} \quad E_i = \eta_E(x_i) \in \mathbb{R}, \tag{7.2}$$

given the system dynamics $\dot{x}_i = f(x_i) + g(x_i)u_i$, where $x \in \mathbb{R}^n$ and $u \in \mathbb{R}^m$, we can still express the constraints in terms of robot positions and energy levels [389, 426].

Robot-Robot Interactions
The first constraint in the survivability catalog is the most primary of safety constraints, namely the assurance that robots do not collide with each other. As this constraint is a *pairwise* constraint, it will prove useful to call out exactly what pairs of agents are being considered. To that end, given N agents, we will continue the practice of letting $[N]$ denote the index set $\{1, \ldots, N\}$, and we associate edge sets with subsets of $[N] \times [N]$ in the sense that the pair $(i, j) \in E$ means that the constraint, defined over the edge set $E \in [N] \times [N]$, pertains to Robots i and j. Collision-avoidance in its most general form involves all pairs of agents, and this must be captured by the corresponding edge set. To this end, let E_{K_N} be the edge set associated with the complete graph, $K_N = ([N], E_{K_N})$, which is the graph without self-loops but where all nodes are pairwise connected, i.e., $(i, j) \in E_{K_N} \Leftrightarrow (i, j) \in [N] \times [N], i \neq j$.[4]

Adhering to Reynolds' Boids terminology, the already encountered *separation* constraint states that

$$h^{pair}_{separation}(x_i, x_j) = \|p_i - p_j\|^2 - D_s^2 \geq 0, \quad (i, j) \in E_{K_N}, \tag{7.3}$$

given some safety distance, $D_s > 0$, and where we continue the practice from the previous chapter of using the superscript *pair* to call out the pairwise nature of the constraint. Or, phrased across all pairs of agents through the Boolean conjunction of constraints (expressed using the min operator),

$$h_{separation}(x) = \min_{(i,j) \in E_{K_N}} \left\{ h^{pair}_{separation}(x_i, x_j) \right\} \geq 0. \tag{7.4}$$

Staying apart is not the only type of relevant pairwise constraint that must be considered. As we have seen, animals and robots alike oftentimes do better in groups. By teaming together, the risk of attack to any one individual can be

---

[4]We saw in Chapter 4 that we actually do not need to consider all pairs, as only nearby agents run the risk of colliding.

mitigated and diluted—similar to W. D. Hamilton's *selfish herd* concept [172]. This is a strategy that, for example, is employed by water skaters (*Halobates robustus*) as they sit together on the water surface, while predator fish snap at them from below [107]. Similarly, foraging and prey capture strategies tend to improve as more individuals participate in a collaborative manner (up to a point), e.g., [99, 107, 175, 180].

Among robots, staying together allows for the establishment and maintenance of a connected interaction network, i.e.,

$$h_{connect}(x) = \min_{(i,j) \in E} \left\{ h^{pair}_{connect}(x_i, x_j) \right\} \geq 0, \tag{7.5}$$

where the pairwise constraints are

$$h^{pair}_{connect}(x_i, x_j) = D_c^2 - \| p_i - p_j \|^2, \ (i, j) \in E, \tag{7.6}$$

given some edge set, E, that is to be maintained, e.g., by the connectivity disk-graph.[5]

In Chapter 3, additional network topology constraints were introduced in order to support richer behaviors that require something stronger than just connectivity. As seen, cyclic pursuit requires a cycle graph, formation control a rigid or shape-similar network, and coverage control a Delaunay graph, just to name a few. To this end, let the network topology required by a particular multi-robot behavior be given by the edge set $E_T \in [N] \times [N]$. The constraint, $h_{topology}$, subsequently takes on a form almost identical to $h_{connect}$, with the only difference being the edge set over which the conjunction is defined, i.e.,

$$h_{topology}(x) = \min_{(i,j) \in E_T} \left\{ h^{pair}_{connect}(x_i, x_j) \right\}. \tag{7.7}$$

Environmental Connections

With $h_{separation}$, $h_{connect}$, and $h_{topology}$ in the books, that takes care of the pairwise interaction constraints. The next category needed to encode robot survival pertains to the connections between robots and their environments. Inspired by spatial distribution concepts such as the ideal free distribution (competition by "exploitation") and the despotic distribution (competition by "resource defense"), e.g., [107, 308, 357], one class of environmental

---

[5]As seen in Chapter 6, this graph need not be static as it is possible to formulate the connectivity constraint as a disjunctive constraint over all feasible, connected networks.

constraint has to concern itself with space usage and the distribution of robots throughout the environment.

In nature, the ideal free distribution is the result of animals spreading out across an area in order to achieve equalized effective resource utilization across the population, which in the robotics domain would correspond to an effective coverage of the area [95, 330, 345]. Similarly, the despotic distribution arises when an animal (or group of animals) defends a region with richer resources, forcing other individuals to settle for less desirable habitats. In that case, the robots would not try to balance and equalize the area-coverage, but rather maximize the resources available to them in their particular region of dominance, which is reminiscent of perimeter and boundary defense strategies employed in some multi-robot applications [84, 175, 377]. Regardless of which distribution concept to draw from, and which robotics interpretation to adhere to, we will stick to the general coverage control framework for the sake of generality as well as for the purpose of aligning with standard robotics nomenclature.

The coverage constraint from Chapters 3 and 4 (Equations 3.12 and 4.28) states that

$$h_{coverage}^{ind}(x_i) = \int_{T_i} Q(\|p_i - q\|)\phi(q)dq - c \geq 0, \tag{7.8}$$

where the superscript *ind* stresses the fact that this is a constraint defined with respect to an individual robot as opposed to the pairwise constraints that involved two robots for their formulation.[6] Additionally, $Q(\|p_i - q\|)$ is a decreasing function, encoding the quality of the measurements obtained at location $q \in T_i$ (or, the ease by which the individual can access the resources located at q), when the robot is positioned at p_i. The domain, T_i, is Robot i's "territory," or region of dominance, and $c > 0$ is the amount of "resources," as interpreted broadly, required by the robot to sustain itself. Note that other variants to this constraint can be conceived and formulated, as was done in both Chapters 2 and 5.

Although an individual robot can enforce the constraint in Equation 7.8 without the need for explicit coordination—see Chapter 4—it is instructive to explicitly write out the team-level coverage constraint using the conjunction across all individuals,

---

[6]Note, if the domain, T_i, is a Voronoi cell, or is based on some other state-dependent tessellation scheme, e.g., [263], then $T_i(x)$ becomes a function of the state. As such, the corresponding barrier function no longer solely depends on x_i, but on the combined state, x.

$$h_{coverage}(x) = \min_{i \in [N]} \left\{ h^{ind}_{coverage}(x_i) \right\} \geq 0. \tag{7.9}$$

Another critical, environmental constraint is to not drive into things in the environment. This can be formulated by a direct modification to the pairwise separation constraint to reflect avoidance of collisions with obstacles in the environment rather than with other robots. Given an obstacle located at $o_j \in \mathbb{R}^d$, together with a suitable separation distance, $D_o > 0$, the constraint becomes

$$h^{ind}_{obstacle}(x_i, o_j) = \| p_i - o_j \|^2 - D_o^2 \geq 0, \ i \in [N], \tag{7.10}$$

or, taken across all known obstacles with index set $[O]$,

$$h^{ind}_{obstacle}(x_i) = \min_{j \in [O]} \left\{ \| p_i - o_j \|^2 - D_o^2 \right\} \geq 0. \tag{7.11}$$

Here, the separation distance, D_o, has to be chosen so as to capture the fact that obstacles are not points, but rather have a non-trivial, spatial footprint. This distance can be selected in a systematic manner without making the response overly conservative, as shown in [72, 83, 159].

The team-level version is given by

$$h_{obstacle}(x) = \min_{i \in [N]} \left\{ h^{ind}_{obstacle}(x_i) \right\} \geq 0, \tag{7.12}$$

even though this constraint can only meaningfully be enforced at the level of individuals (Equation 7.10), rather than as a team.

Rounding out the group of environmental constraints is the *spatial anchoring* constraint, discussed in Chapter 4, whereby a robot's motion is constrained to ensure that it always has enough charge left in the battery to make it back to a select location. For an animal, such spatial anchoring would arise around critical resources (e.g., particular diet items or water sources), that are only available at select locations [378]. For robots, these locations would typically be charging stations or communication hubs [302].

The spatial anchoring constraint, as per Equation 4.29, states that

$$h^{ind}_{anchor}(x_i) = (E_i - E_{min}) - \mathcal{E}(dist(p_i)) \geq 0, \tag{7.13}$$

where $dist(p_i)$ is the distance from the robot's position to the anchor location, $\mathcal{E}(dist(p_i))$ is the energy required to travel that distance under suitable assumptions on the motion model, e.g., [374], and E_{min} is the smallest acceptable energy level. Following the conjunctive procedure for the team-level constraint yields

$$h_{anchor}(x) = \min_{i \in [N]} \left\{ h^{ind}_{anchor}(x_i) \right\} \geq 0. \tag{7.14}$$

Internal Conditions/Homeostasis
The reason why the constraint in Equation 7.14 is different from the homeostasis constraint, introduced in Chapter 3 and elaborated on further in Chapter 5, is that it is ultimately a statement about movement through the environment, as opposed to a statement about internal conditions. In fact, the last among the required, constituent survivability constraints is exactly driven by this need for an internally focused constraint (i.e., homeostasis—to borrow the ecological term). Arguably, the entire focus of Chapter 5 was on this singular constraint in order to achieve task persistification.

The "homeostasis" constraint states that

$$E_i \in [E_{min}, E_{max}], \tag{7.15}$$

where E_{min} and E_{max} are the minimum and maximum energy levels supported by Robot i's battery configuration. The maximum value is quite important as it is what will prevent robots from simply staying at charging stations indefinitely through the coupling with the environment, i.e., the robot will eventually be forced away from the charging station in order not to over-charge.[7] The corresponding barrier function becomes

$$h^{ind}_{energy}(x_i) = (E_{max} - E_i)(E_i - E_{min}) \geq 0, \tag{7.16}$$

or, taken across the entire robot team,

$$h_{energy}(x) = \min_{i \in [N]} \left\{ h^{ind}_{energy}(x_i) \right\} \geq 0. \tag{7.17}$$

What we have now done is cataloged seven constraints, pertaining to the three broad themes of *robot-robot interactions*, *connections between robot and its environment*, and *internal conditions (homeostasis)*. These are summarized as follows:

---

[7]The prevention of over-charging was ensured in a slightly different (and elaborate) manner in Chapter 4, but thanks to the high relative degree formalism introduced in Chapter 5, the more simple and direct form in Equation 7.15 suffices.

Interactions	Environment	Homeostasis
$h_{separation}$	$h_{coverage}$	h_{energy}
$h_{connect}$	$h_{obstacle}$	
$h_{topology}$	h_{anchor}	

Note that it is possible to imagine more such constituent constraints. Or to encounter situations where not all seven are relevant. As such, this should not be interpreted as the definitive list of relevant survivability constraints. Rather, it is a sampler of what such constraints could (and perhaps should) look like. More importantly, it provides us with enough expressiveness to be able to formulate more complex constraints in order to articulate the robot ecology framework, culminating with the composite survivability constraints themselves, as discussed in the next section.

7.1.2 Survivability Constraints

At the very core of the robot ecology story is a robot, expending as little energy as possible, simply being present and sustaining itself in the environment, and preparing to be recruited and tasked with carrying out new and previously unencountered missions. Expending as little energy as possible is a clear concept—the cost to be minimized is the energy, i.e., the control input should be chosen to satisfy min $\|u\|^2$ (subject to constraints), e.g., [66, 238]. But what exactly does *surviving* entail?

With the constituent constraints established, we are now within striking distance of formulating such survivability constraints, while the question of taskable robots will have to be deferred to the next section. Before we begin, a few words need be said about *survivability* versus *survival*, which we have sometimes used almost interchangeably throughout the book. "Survival" simply means that the robots never experience any deployment-ending conditions (such as getting stuck, running out of battery energy without the ability to recharge, colliding with obstacles in a severely damaging way, etc.), while "survivability" refers to an ability to survive despite being asked to do a number of other things as well. For that reason, we prefer to use the term survivability as it captures the idea of a robot being recruited and tasked with carrying out a diverse set of missions, all the while guaranteeing that its ability to survive (survivability) is not compromised.

We have previously expressed survivability through the colloquial "Don't die!" statement. But what is hiding behind those words could be something much more elaborate, like "Stay away from predators AND have enough energy to always be able to replenish the energy sources AND (huddle together for warmth OR (spread out to detect intruders AND form protective formations around the territory))."[8] For robots on long-duration deployments, one can replace "predators" with "malicious agents" and "huddle together for warmth" with "establish a connected information-exchange network," and so forth. Regardless of domain, the combinations one can envision here are significant and diverse, and we build up towards the final survivability constraints by first considering a collection of related, complex yet subordinate constraints.

Foraging

Foraging, i.e., the exploration of an area for the purpose of finding "resources" (as interpreted broadly), is one useful behavior class that can be readily derived from, and grounded in, ecological constraints. Simply put, an animal (or robot) has to acquire the necessary energy and nutrition in order to survive. For a robot (or group of robots) deployed in an environment, foraging would contain elements of recharging/refueling, i.e., the barrier function h_{energy} could capture this need.

The foraging concept has additional implications for an organism's space use and selection of particular habitat types [308, 332]. In other words, an organism must forage across a sufficient area to support its energetic needs, i.e., $h_{coverage}$ may be of relevance as well. Finally, if critical resources, such as charging stations, are only available at select locations, the organism is spatially anchored, and is forced to return to those locations, i.e., h_{anchor} may be needed. As a result, the foraging constraint could look something like

$$h_{forage}(x) = h_{anchor}(x) \wedge (h_{coverage}(x) \vee h_{energy}(x)), \qquad (7.18)$$

i.e., the animal/robot should always have enough energy to return to the spatially anchored feeding/charging locations. Additionally, it should either be covering a sufficiently large area, or be engaged in the process of charging, through the energy constraint.[9]

---

[8] We capitalize the Boolean operators to explicitly call out the fact that survivability, by necessity, is a composite constraint.

[9] Note here that we are using the established notation, $h = h_1 \wedge h_2$, as shorthand for the Boolean proposition $h \geq 0 \Leftrightarrow (h_1 \geq 0 \wedge h_2 \geq 0)$, which, in turn, can be stated as the constraint $h = \min\{h_1, h_2\} \geq 0$.

In Chapter 3, the constraint sets $\mathscr{G}_{resources}$ and $\mathscr{G}_{homeostasis}$ were introduced and the proposed formulation of foraging as a behavior resulting from a number of constraints is in fact, after untangling the constraint sets and the barrier function in Equation 7.18, revealed to be equivalent to ensuring that the state of the system belongs to the intersection, $\mathscr{G}_{resources} \cap \mathscr{G}_{homeostasis}$. Such ecological connections are certainly not coincidences. Foraging, in behavioral ecology, can be characterized as being oriented towards the maximum net rate of delivery of food, e.g., by birds returning to their nest with food for their chicks as effectively as possible. The nest site would serve as the anchor, and the search for food would be captured through the coverage constraint.

Traveling back and forth between the nest and the resource site takes time and energy, and the search for food is not free either. In fact, the search for food typically has a diminishing return in that the more food already found in an area, the longer it takes to find additional food sources. A similar conundrum is faced by honey bees. But, rather than an increased time between food discoveries, their diminishing return stems from the fact that the weight of the nectar has a significant and nonlinearly increasing energetic effect on the flight back to the hive [107]. (This model of load carrying and diminishing return within a patch is known as the *Marginal Value Theorem* [80].)

The same kind of qualitative foraging decisions faced by birds and bees must also be made by predators when it comes to prey selection. If the energy value of Prey i is ϵ_i, and its "handling time" is τ_i, then the profitability associated with consuming the prey is ϵ_i/τ_i. But, time searching for the prey, σ_i, matters as well. So, if a predator has encountered Prey i, but is pondering passing up this meal and instead to go looking for Prey j, this is the right choice if

$$\frac{\epsilon_j}{\tau_j + \sigma_j} > \frac{\epsilon_i}{\tau_i}. \tag{7.19}$$

This type of economic foraging calculus is implicitly captured by the constraint in Equation 7.18 through the interplay between the constituent *coverage* (search for food), *energy* (consume the food), and *anchor* (return the food to a particular location) constraints.

Interactions and Territoriality

As seen in Chapter 3, competitive *interactions* and *territoriality* within species give rise to fundamental, ecological constraints. Within many species, individuals avoid one another to limit competition, and to limit "interference" with one another. For robots deployed over long time-scales, "territoriality" could correspond to robots not getting too close, encoded through the barrier function $h_{separation}$. When combined with an information exchange network requirement (thereby ensuring cohesion, to borrow, yet again, from Reynolds' playbook) through $h_{connect}$, one could produce the following "proto-territoriality," interaction constraint, encountered already under the moniker $h_{navigation}$ in Equation 6.46,

$$h_{network}(x) = h_{separation}(x) \wedge h_{connect}(x), \tag{7.20}$$

or the slightly stronger and more involved,

$$h_{network}(x) = h_{separation}(x) \wedge h_{connect}(x) \wedge h_{topology}(x), \tag{7.21}$$

to align better with the ecological interaction constraint set, $\mathcal{G}_{interact}$, from Chapter 3.

A variation on this theme could be to combine the interaction constraint with the foraging constraint to arrive at a situation where the robots spread out and cover an area while, at the same time, avoid collisions and, more importantly, are able to return to a charging station to recharge their batteries. The behavior resulting from this combination of the foraging and interaction constraints approaches something akin to the territorial behaviors found in nature, i.e.,

$$h_{territory}(x) = h_{forage}(x) \wedge h_{network}(x). \tag{7.22}$$

Survival

Having seen a few examples of complex behaviors inspired by behavioral ecology that can be produced through Boolean compositions of constituent constraints, we now return to the concluding observation about the survivability set in Chapter 3, namely that the state of the system should remain in the following set for all times,

$$\mathcal{G}_{survive} = \mathcal{G}_{resources} \cap \mathcal{G}_{interact} \cap \mathcal{G}_{homeostasis}. \tag{7.23}$$

These three sets, as depicted in Figure 7.1, line up well with the designation of constituent constraints as either dealing with environmental connections (resources), interactions, or internal conditions (homeostasis).

As already hinted at, survivability can come in different varieties depending on what types of robots and deployments are being considered. We here describe how a few different situations map to $\mathcal{G}_{survive}$, through the use of the

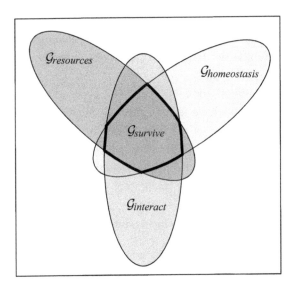

Figure 7.1: Survival, viewed as an intersection of three sets, pertaining to environmental connections ($\mathcal{G}_{resources}$), robot-robot interactions ($\mathcal{G}_{interact}$), and internal conditions ($\mathcal{G}_{homeostasis}$).

aforementioned, subordinate barrier functions. But, regardless of what barrier functions are being used, the survivability constraint will, in light of Equation 7.23, adhere to the unified form,

$$h_{survive}(x) = h_{resources}(x) \wedge h_{interact}(x) \wedge h_{homeostasis}(x). \tag{7.24}$$

Since not all of the constituent sets in Equation 7.23 are required in all scenarios, it is useful to introduce the dummy, identity constraint, $h_{identity}(x) = 1$. This constraint trivially satisfies $h_{identity} > 0$, i.e., it never has any effect on the actions the robots may take. But it allows us to conform to the general expression in Equation 7.24.

No matter the situation, not slamming into things in the environment and not completely depleting the batteries are always of importance, i.e., the most rudimentary of survivability constraints would consist of $h_{resources} = h_{obstacle}$, $h_{interact} = h_{identity}$, $h_{homeostasis} = h_{energy}$. If the deployment involves more than one robot, then the interaction constraint would have to be modified to include inter-robot collision-avoidance, i.e., $h_{interact} = h_{separation}$. Along these lines, if what the robots have to achieve is not only separation, but also to establish a connected interaction network, we get $h_{interact} = h_{separation} \wedge h_{connect}$. Additionally, if what is required is the maintenance of a particular network topology, the interaction constraint expands out to $h_{interact} = h_{separation} \wedge h_{connect} \wedge h_{topology}$.

Another differentiator when it comes to the types of constraints to use is whether or not the robots have to return to a dedicated location to recharge, or if this is left to the robots to "figure out" through the sole application of the homeostasis constraint. If they are indeed spatially anchored, the resource constraint can no longer only involve avoidance of obstacles in the environment, but must include the anchor constraint as well, i.e., $h_{resources} = h_{obstacle} \wedge h_{anchor}$.

We summarize these observations in the following table:

$h_{survive}(x) = h_{resources}(x) \wedge h_{interact}(x) \wedge h_{homeostasis}(x)$	
Constituent Constraints	Characterization
$h_{homeostasis} = h_{energy}$	Task persistification must always hold
$h_{resources} = h_{obstacle}$	Avoiding collisions with obstacles
$h_{resources} = h_{obstacle} \wedge h_{anchor}$	Robots are spatially anchored
$h_{interact} = h_{identity}$	Single robot deployment
$h_{interact} = h_{separation}$	Multiple robots, no connectivity requirements
$h_{interact} = h_{separation} \wedge h_{connect}$	Robot network should always be connected
$h_{interact} = h_{separation} \wedge h_{connect} \wedge h_{topology}$	A particular network topology is required

Table 7.1: Constraints for ensuring robot survival.

The one constraint that does not quite fit neatly into this organizational structure is the coverage constraint. In the foraging formulation, coverage had to be coupled disjunctively with the energy constraint—either the robot is covering an area *or* it is charging at a charging station—which is *not* consistent with the unified survivability formulation in Equation 7.24. The core of this conceptual mismatch is that it is conceivable that the environment simply does not support the requisite resource "mass" in each domain, as encoded through the density function and the cutoff for acceptable coverage in Equation 7.8. In fact, if these parameters are not selected carefully, it is entirely possible that the robots would not be able to cover enough mass. Or, even if they could, they may simply linger forever near the charging stations since that way, $h_{energy} \geq 0$ while $h_{coverage} < 0$ for all times, thereby rendering the disjunction in h_{forage} true for all times.

One can resolve this by a careful tweaking of the relevant parameters, as was done in [128]. But, that does not seem like the right way to go if one wants to produce robot systems that can be deployed *robustly* (thereby maximizing the fitness set) over long periods of time. In fact, any strategy that hinges on carefully selected parameters is almost certainly doomed to fail in the context of long-duration autonomy—*what can go wrong will sooner or later go wrong*. For this reason, we exclude coverage from the survivability constraints and instead postpone the discussion about coverage to the next section, which will deal with the issue of goal-driven behaviors.

7.2 Goal-Driven Behaviors

As we just noted, there are behaviors and objectives that do not readily fit the constraint-based survivability formalism in Equation 7.24. It may or may not be possible to always guarantee that there are sufficient resources in a particular area, as the environment may simply not be resource rich enough. To this end, goal-driven *maximization* of resources seems like a natural, next best option. Correspondingly, in nature, central to individual fitness is *reproductive success* (and the associated *mating ecology*), whereby certain behaviors occupy a special place in the behavioral taxonomies in that an individual's reproductive success could be viewed as an *objective*, or *goal*, rather than as a constraint, e.g., [7]. For such reasons, the constraint-based survivability paradigm needs to be augmented to support goal-driven behaviors as well.

To put this required modification in its proper context, consider two slightly different yet equally important aspects of the *robot ecology* idea, namely that (*i*) survivability should be encoded as a constraint, which can be achieved through Boolean composition of multiple constituent barrier functions, and (*ii*) the robots should be recruitable and taskable to carry out user-specified instructions, tasks, and even full-blown missions, in a manner that does not break the structure of the survivability constraints.

The way these two aspects have been approached previously, in Part II, is by following a constrained optimization semantics, where the idea is to let a nominal control input, u_{nom}, correspond to the user specifications. Subsequently, the actual control input, u, is taken to be the solution to the least-squares problem,

$$\min_{u} \|u - u_{nom}\|^2, \qquad (7.25)$$

subject to the survivability CBF constraint

$$S_{h_{survive}}[x; u] \geq 0, \tag{7.26}$$

or, expanded out,

$$L_f h_{survive}(x) + L_g h_{survive}(x)u \geq -\alpha(h_{survive}(x)), \tag{7.27}$$

as per Equation 4.17. Here we have assumed that the robot dynamics is given by a control-affine system, $\dot{x} = f(x) + g(x)u$, and where we have fudged the fact that the Lie derivates may or may not be set-valued, as per Chapter 6.

This certainly seems like a natural way of injecting external commands for the purpose of recruiting and tasking robots. But, although both elegant and useful (and used) in a number of applications, such as adaptive cruise-controllers [13], bipedal humanoid robots [297, 354], swarming quadcopters [426], and teleoperation [300], there is something slightly unsatisfactory with this formulation from a robot ecology vantage point. Based on the ecological observation about a "purposeful expenditure of energy," energy-minimization and power-management have played key roles in the story thus far. As such, what the robot ecology framework calls for is that the cost to be minimized really should be the straight-up energy cost, i.e., the control input should be given by

$$\min_u \|u\|^2, \tag{7.28}$$

subject to the survivability constraint in Equation 7.24. But then we lose the recruitability feature that we just insisted on. What is needed is some way of embracing both of these seemingly conflicting requirements —*minimized energy* and *constraints rather than goals as the primary driver of behavior*— and in this section, we show how this is indeed possible to accomplish.

As u_{nom} is typically expected to come from some exogenous, performance-related optimization problem that encodes the task, expressed by the cost $J_{task}(x)$, what we will do is turn this cost into a constraint in a systematic manner. Once that has been accomplished, it is tempting to target a Boolean conjunction of the survivability constraint with the task constraint, i.e., to aim for something along the lines of $h_{ecology} = h_{survive} \wedge h_{task}$, encoded through the min operator, $h_{ecology}(x) = \min\{h_{survive}(x), h_{task}(x)\}$.

Unfortunately, as will be seen, we will not be able to fully realize this ambition to turn the cost into a constraint to be rendered forward invariant. In fact, we will end up with a task constraint that is negative rather than positive, and that asymptotically approaches some desired target value. The questions then become those of figuring out how exactly this constraint should be formulated, and how it can be systematically combined with the survivability constraints.

7.2.1 From Gradient Descent to Barrier-Based Descent

To get at the issue of how to incorporate goal-driven performance considerations within the minimum energy framework, let us return to the way multi-robot controllers were constructed in Chapter 2. There, a performance cost was formulated in terms of the joint robot states, and the control law was obtained by flowing against the gradient of this cost.

To elucidate this point, let $J : \mathbb{R}^n \to \mathbb{R}_{\geq 0}$ be a continuously differentiable, positive definite performance cost, with $J(x^\star) = 0$ at a minimizer, $x^\star$, and $J(x) > 0$ otherwise.[10] The Chain Rule dictates that, along trajectories,

$$\frac{d}{dt} J(x(t)) = \frac{\partial J}{\partial x}^T \dot{x}. \tag{7.29}$$

If $\dot{x} = f(x) + g(x)u$, and if we were able to select the control input, u, in such a way that

$$f(x) + g(x)u = -\nabla J(x), \tag{7.30}$$

we would get

$$\frac{d}{dt} J(x(t)) = -\frac{\partial J}{\partial x}^T \frac{\partial J}{\partial x} = -\left\| \frac{\partial J}{\partial x} \right\|^2 \leq 0, \tag{7.31}$$

as shown in Figure 7.2 (Top).

By LaSalle's Invariance Principle, this continuous-time gradient descent algorithm will drive the system to a point where the derivative vanishes, which, in turn, is a stationary point to the minimization problem. As such, in a certain sense, the problem is "solved." As an example, if $\dot{x} = u$, and

$$u = -\frac{\partial J}{\partial x}, \tag{7.32}$$

the state will indeed converge asymptotically to a local minimizer.

---

[10] Since the forthcoming discussion does not require an external "task" interpretation, we use $J(x)$ rather than $J_{task}(x)$ to keep the exposition general.

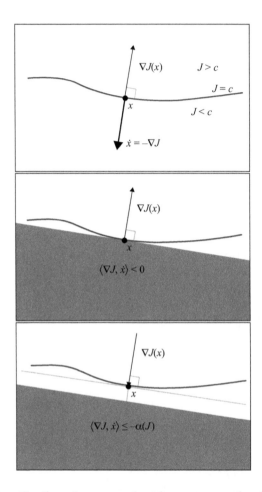

Figure 7.2: Top: Gradient descent makes the system evolve in the direction where the cost is reduced the most (locally). Middle: An entire open half-space (shaded) of directions is available that will all result in a (local) reduction of the cost. Bottom: By introducing an offset, the shifted, closed half-space contains directions where the local reduction is bounded away from zero.

But, the choice of control action in Equation 7.32, or more generally, to insist on $\dot{x} = -\nabla J(x)$, is a rather specific and extreme (it is called "*steepest* descent" after all) choice. It certainly does not leave a lot of room for flexibility in terms of minimizing the cost, while simultaneously satisfying survivability constraints. But there really is no need to flow in exactly this direction. In fact, consider the Taylor expansion around the point x,

$$J(x + \delta x) = J(x) + \frac{\partial J(x)}{\partial x}^T \delta x + o(\|\delta x\|). \tag{7.33}$$

For small δx, we have that

$$J(x + \delta x) - J(x) \approx \langle \nabla J(x), \delta x \rangle, \tag{7.34}$$

which is negative (meaning that the cost is reduced), as long as the inner product satisfies

$$\langle \nabla J(x), \delta x \rangle < 0. \tag{7.35}$$

Now, assuming that the point around which the expansions is taken, x, is actually $x(t)$ at some given time t, then $\delta x(t) \approx \dot{x}(t) \delta t$, for some short time horizon, $\delta t > 0$. As a result, the descent condition translates to the constraint

$$\langle \nabla J(x), \dot{x} \rangle < 0, \tag{7.36}$$

as shown in Figure 7.2 (Middle). Although not quite a CBF constraint, the inequality in Equation 7.36 sure does seem highly related to one. And this is something that we can potentially leverage when turning costs into constraints. But for that, we have to make it resemble a CBF constraint even more.

The first order of business is to turn the strict inequality in Equation 7.36 into a non-strict one. The way to do that, while at the same time providing an additional descent "cushion," could be to insist on

$$\langle \nabla J(x), \dot{x} \rangle \leq -\alpha(J(x)), \tag{7.37}$$

for some locally Lipschitz continuous, extended class $\mathcal{K}_\infty$-function, α, as illustrated in Figure 7.2 (Bottom). This takes us even closer to a CBF. But, we are not there quite yet. The source of this slight inconsistency is that the inequality sign in Equation 7.36 is reversed, as compared to the regular CBF constraint.

To get at the required reversal of the inequality constraint, consider the function $h_J(x) = -J(x)$. It is zero at minimizers to $J(x)$, and strictly negative elsewhere. Swapping these functions turns the inequality in Equation 7.36 into

$$\langle -\nabla h_J(x), \dot{x} \rangle \leq -\alpha(-h_J(x)), \tag{7.38}$$

i.e.,

$$\langle \nabla h_J(x), \dot{x} \rangle \geq \alpha(-h_J(x)). \tag{7.39}$$

One of the properties required by the extended class $\mathcal{K}_\infty$-functions is that they are increasing functions passing through the origin. As such, we can define another extended class $\mathcal{K}_\infty$-function, $\hat{\alpha}$, through

$$\hat{\alpha}(z) = -\alpha(-z), \tag{7.40}$$

which retains all required properties needed for a function to be extended class $\mathcal{K}_\infty$.[11]

With this choice of extended class $\mathcal{K}_\infty$-function, the inequality in Equation 7.39 becomes

$$\langle \nabla h_J(x), \dot{x} \rangle \geq -\hat{\alpha}(h_J(x)). \tag{7.41}$$

Now, this is a proper CBF constraint! The only dissonant note is that rendering $h_J(x) \geq 0$ forward invariant is mostly impossible. Unless x is a minimizer to $J(x)$, we have that $h_J(x) < 0$, since $J(x)$ is positive definite and $h_J(x) = -J(x)$.

Luckily for us, by returning to the formulation in Equation 7.37, all the tools needed to manage the situation are actually already provided. Although this constraint is not a CBF constraint, it is something equally powerful, namely a *Control Lyapunov Function* (CLF) constraint, e.g., [209]. Staying with the notation established thus far, and relying on the particular formulation in [11], the following key result is highly useful.

Theorem 7.1 Given the control-affine, dynamical system, $\dot{x} = f(x) + g(x)u$, with f and g being locally Lipschitz continuous, and a continuously differentiable, positive definite function, $J : \mathbb{R}^n \to \mathbb{R}$, with $J(0) = 0$. Any continuous controller, u, such that

$$L_f J(x) + L_g J(x)u \leq -\alpha(J(x)), \tag{7.42}$$

for some extended class $\mathcal{K}_\infty$-function, α, renders the origin, $x = 0$, asymptotically stable.

---

[11] Since we do not require that α be antisymmetric, we do not, in general, have $\hat{\alpha} = \alpha$ in Equation 7.40.

Since Equations 7.37 and 7.41 are equivalent, what we have actually done is recovered the robustness property associated with CBFs in that even if the system starts outside the super-level set, $\{x \mid h(x) \geq 0\}$, it will asymptotically converge to that set [436]. Or, following [306] and phrased in terms of the minimization of a particular, positive definite performance cost, $J(x)$, what we have done is discovered the following handy result.

Theorem 7.2 Given the system, $\dot{x} = f(x) + g(x)u$, and performance cost, $J(x)$, as per Theorem 7.1. Set $h_J(x) = -J(x)$, and let u satisfy the inequality constraint

$$L_f h_J(x) + L_g h_J(x)u \geq -\alpha(h_J(x)), \tag{7.43}$$

for some extended class $\mathcal{K}_\infty$-function, α. Then, if the system starts away from $J(x(t_0)) = 0$, it will asymptotically approach a minimizer to $J(x)$, i.e., $\lim_{t \to \infty} x(t) \in \{x \mid J(x) = 0\}$.

Of course, the set $\{x \mid J(x) = 0\}$ is still rendered forward invariant under Equation 7.41. As the only point(s) where $J = 0$ are the minimizers to the cost, what this means is that if one were to miraculously happen to start out at a minimizer, one would remain at a minimizer. But, we certainly cannot count on being that lucky. Regardless of this oddity, Theorem 7.2 gives us all the tools needed to turn costs into constraints for the purpose of long-duration autonomy and robot ecology.

7.2.2 Costs as Constraints

The intriguing connection between CBFs and CLFs seems quite promising.[12] In fact, the combination of CLFs for performance with CBFs for safety is a powerful duo that has been leveraged in a number of contexts, e.g., [11, 391]. For the purpose of robot ecology, however, it is not quite perfect just yet. The reason for this is the slight discomfort caused by the fact that if the CLF is equal to the performance cost, then the system will indeed be driven (asymptotically) to a minimizer to J, as per Theorem 7.2. But, as established, optimality is not only not needed, it is actually not even all that desirable from

---

[12] The standard notation is to use V, rather than J, to denote the Lyapunov function, eg., [209].

a robustness point of view [120]. In other words, we simply want solutions that are somehow "good enough."

In order to explicitly retain the performance-based considerations provided by the CLFs, without overly fussing over optimality, one can introduce a slack variable, δ, that encodes this idea of "good enough" performance. To this end, consider the following constrained optimization problem.

$$\min_{u,\delta} \|u\|^2 + \delta^2 \qquad (7.44)$$

subject to the constraint

$$L_f h_J(x) + L_g h_J(x)u \geq -\alpha(h_J(x)) - \delta. \qquad (7.45)$$

This formulation retains the desired minimum energy flavor, yet allows for an explicit tradeoff between energy (captured by $\|u\|^2$) and performance (as encoded by δ^2). And, to get a better feeling for why this construction works, consider the drift-free system,

$$\dot{x} = g(x)u, \qquad (7.46)$$

i.e., where the absence of a drift term ensures that the system simply stops evolving whenever $u = 0$.

In the drift-free case, the introduction of the slack variable, δ, does not matter (asymptotically), in the absence of other constraints in that the minimizer to the original, slack-free minimum-energy problem is recovered. To see this, consider the KKT conditions to the problem in Equations 7.44 and 7.45,

$$-\frac{\partial h(x)^T}{\partial x} g(x)u^* - \alpha(h(x)) - \delta^* \leq 0 \qquad (i)$$

$$\lambda^* \geq 0 \qquad (ii)$$

$$\lambda^* \left(-\frac{\partial h(x)^T}{\partial x} g(x)u^* - \alpha(h(x)) - \delta^*\right) = 0 \qquad (iii) \qquad (7.47)$$

$$2u^* - g(x)^T \frac{\partial h(x)}{\partial x} \lambda^* = 0 \qquad (iv)$$

$$2\delta^* - \lambda^* = 0 \qquad (v)$$

where u^*, δ^* and λ^* are primal and dual optimal points, e.g., [58].

If $\lambda^* = 0$ then, by (iv) and (v), $u^* = 0$ and $\delta^* = 0$. This means that $\dot{h}(x) = 0 \geq -\alpha(h(x))$, per (i). But, if this does not correspond to a minimizer, then $h(x) < 0$ (since $J(x) > 0$). As such, $\alpha(h(x)) < 0$ as well, which means that

$\dot{h}(x) > 0$, which is a contradiction. The conclusion that must be drawn is thus that this situation corresponds to a minimizer to $J(x)$.

On the other hand, if $\lambda^* > 0$, then

$$u^* = \frac{1}{2}g(x)^T \nabla h(x)\lambda^* \text{ and } \delta^* = \frac{1}{2}\lambda^*, \tag{7.48}$$

as per (iv) and (v). Plugging this into (iii) yields,

$$-\frac{1}{2}\nabla h(x)^T g(x)g(x)^T \nabla h(x)\lambda^* - \alpha(h(x)) - \frac{1}{2}\lambda^* = 0, \tag{7.49}$$

i.e.,

$$\lambda^* = -2\frac{\alpha(h(x))}{1 + \|g(x)^T \nabla h(x)\|^2}$$

$$\text{and } u^* = -\frac{\alpha(h(x))}{1 + \|g(x)^T \nabla h(x)\|^2}g(x)^T \nabla h(x). \tag{7.50}$$

Using this in the expression for $\dot{h}(x)$ results in

$$\dot{h}(x) = -\frac{\alpha(h(x))\|g(x)^T \nabla h(x)\|^2}{1 + \|g(x)^T \nabla h(x)\|^2}. \tag{7.51}$$

Since $\alpha(h(x)) < 0$, away from a minimizer, if $\nabla h(x) \neq 0$, then $\dot{h}(x) > 0$. On the other hand, if $\nabla h(x) = 0$ then $\dot{h}(x) = 0$ as well. What this means, thanks to LaSalle's Invariance Principle, is that, as $t \to \infty$, $x(t) \to x^*$ such that $\frac{\partial J}{\partial x}(x^*) = 0$. As a result, $J(x^*) = 0$, which implies that this is indeed a minimizer, as per [306]. The interpretation here is that the introduction of the slack term is entirely harmless—in fact, even borderline pointless—for drift-free systems in the absence of additional constraints.

If there is a drift-term present in the dynamics, i.e., $\dot{x} = f(x) + g(x)u$, this asymptotic equivalence may no longer hold. It may simply not be worth the energy investment to drive the state all the way to the minimizer, i.e., driving δ to zero may require too much control effort. Similarly, if there are other constraints present, which there undoubtedly will be, finding such minimizers is no longer even something to strive for. Good enough will have to be, well, good enough.

After that slack variable detour, we now return to the originators of this discussion, namely a task specified through a cost, $J_{task}(x)$, and a desire to treat this cost as a constraint in a manner that combines with survivability in a natural manner. The necessary ingredients for the production of such a unified formulation have actually already been assembled. What previously was phrased as

$$
\min_u \|u - u_{task}\|^2 \\
\text{subject to } S_{h_{survive}}[x; u] \geq 0 \tag{7.52}
$$

can now be replaced by the robot ecology problem below.

$$
\min_{u, \delta} \|u\|^2 + \delta^2 \\
\text{subject to } \begin{cases} S_{h_{task}}[x; u] \geq -\delta \\ S_{h_{survive}}[x; u] \geq 0. \end{cases} \tag{7.53}
$$

Note here that $h_{task}(x)$ (previously known as $h_{J_{task}}(x)$) is never positive, which is why the Boolean conjunction, formulated in terms of the minimum over the arguments, is not applicable—positivity of $h_{survive}$ and h_{task} cannot be rendered forward invariant. It should also be pointed out that there are no reasons why the same extended class $\mathcal{K}_\infty$-function should be used in the two constraints, $S_{h_{survive}}[x; u]$ and $S_{h_{task}}[x; u]$, even though the formalism in Equation 7.53 suppresses the explicit dependence on the choice of extended class $\mathcal{K}_\infty$-functions.

Before seeing this goal-driven machinery in action, we first introduce a slight variation on the theme, namely a finite-time formulation that highlights why the use of different extended class $\mathcal{K}_\infty$-functions in Equation 7.53 may come in handy. The reason for this is that sometimes the task must be accomplished in a finite amount of time. To that end, the asymptotic quality in Theorem 7.2 may just not be enough.

7.2.3 Finite-Time Performance

The key to enabling the construction of costs as constraints is provided by the control Lyapunov function machinery, as the CLFs act as intermediaries towards the formulation in Equation 7.53. In particular, the asymptotic convergence to the origin in Theorem 7.1 could be transformed into a robustness property for CBFs, in that the system will asymptotically reach the safe set if it starts outside of this set. It is, however, possible to replace the extended class $\mathcal{K}_\infty$-function in Equation 7.37 with a more specialized such function that guarantees that the convergence happens in finite time, rather than asymptotically. The following theorem from [49] makes this point.

Theorem 7.3 Given the setup in Theorem 7.1. Any continuous controller, u, such that

$$L_f J(x) + L_g J(x)u \leq -cJ(x)^\gamma, \tag{7.54}$$

where $c > 0$ and $\gamma \in (0, 1)$, renders the origin, $x = 0$, finite-time stable. Moreover, an upper bound for the settling time, T, is given by

$$T \leq \frac{J(x_0)^{1-\gamma}}{c(1-\gamma)}, \tag{7.55}$$

where x_0 is the value of $x(t)$ at time $t = 0$.

We can leverage this result in a direct manner, but retooled for CBFs rather than CLFs, to get at the desired finite-time property. For example, the discussion of drift-free dynamical systems transfers immediately to the finite-time setting as well [306].

Corollary 7.1 Given the drift-free dynamical system $\dot{x} = g(x)u$, and the objective of minimizing a continuously differentiable, positive definite cost function, $J(x)$. The solution to the optimization problem

$$\min_{u,\delta} \|u\|^2 + \delta^2, \quad \text{subject to} \quad \frac{\partial h_J^T}{\partial x} g(x)u \geq -ch_J(x)^\gamma - \delta, \tag{7.56}$$

where $h_J(x) = -J(x)$, $c > 0$, and $\gamma \in (0, 1)$, will drive the state, x, to a stationary point to the cost, J, in finite time.

Perhaps more importantly, the finite-time variation allows us to let the crucial, forward invariance property of the survivability constraint in Equation 7.53 retain its standard form,

$$S_{h_{survive}}[x; u] = L_f h_{survive}(x) + L_g h_{survive}(x)u + \alpha(h_{survive}(x)) \geq 0, \tag{7.57}$$

while making the task-based constraint be formulated so as to "accomplish" the task in finite time,

$$S_{h_{task}}[x; u] = L_f h_{task}(x) + L_g h_{task}(x)u + ch_{task}(x)^\gamma. \tag{7.58}$$

As we will see in the next section, this ability to achieve objectives in finite time will, for example, come in handy when a team of robots are supposed to toggle through a series of interaction topologies in order to execute a complex, sequential multi-robot maneuver.

7.3   Goal-Driven Multi-Robot Systems

As established in Chapter 2, multi-robot behaviors should be *local, scalable, safe,* and *emergent,* e.g., [93, 253], for them to be viable. A number of behaviors satisfying these properties have been introduced in this book already, with target applications ranging from social dynamics [356], formation control [127], and area coverage [95], just to name a few. In this section, we apply the results derived in the previous section, with the aim of revisiting these "classic" multi-robot behaviors through the new lens of constraint-based design.

Consider, yet again, a collection of N robots, with positions $x_i \in \mathbb{R}^d$, $i \in [N]$, where $d = 2$ for planar robots, and $d = 3$ in the case of aerial robots. Assume each robot is equipped with an omni-directional range sensor that allows it to measure the relative positions of neighboring robots, i.e., Robot i is able to measure $x_j - x_i$, when it is adjacent to Robot j. These interactions among robots are described by the graph $G = (\mathcal{V}, E)$, where $\mathcal{V} = [N]$ is the robot index set, and $E \subseteq \mathcal{V} \times \mathcal{V}$ is the set of edges between robots, encoding the adjacency relationships. If $(i, j) \in E$, then Robot i can measure Robot j's relative position, $x_j - x_i$. We assume that the graph is undirected, i.e., $(i, j) \in E \Leftrightarrow (j, i) \in E$. And, in order to obtain decentralized algorithms, we restrict each robot to act solely based on the locally available information, by which we mean the relative positions of its neighbors. By construction, this leads to inherently scalable, coordinated control algorithms.

As seen in Chapter 2, one systematic way of achieving the desired team-level properties in a decentralized manner is to formulate the performance cost

$$J(x) = \frac{1}{2} \sum_{i=1}^{N} \sum_{j \in \mathcal{N}_i} J_{ij}(\|x_i - x_j\|), \tag{7.59}$$

where $x = [x_1^T, \dots, x_N^T]^T \in \mathbb{R}^{Nd}$, $\mathcal{N}_i$ is Robot i's neighborhood set, and J_{ij} is a symmetric, pairwise cost between Robots i and j, in the sense that

$J_{ij}(\|x_i - x_j\|) = J_{ji}(\|x_j - x_i\|)$. Additionally, to make this formulation fit the recently developed framework further, assume that $J_{ij}(z) \geq 0$, $\forall (i, j) \in E$, $\forall z \in \mathbb{R}_{\geq 0}$, which, in turn, means that $J(x) \geq 0$, $\forall x \in \mathbb{R}^{Nd}$.

Assuming that we can directly control the velocity of Robot i through $\dot{x}_i = u_i$, one can employ the gradient descent flow policy in Equation 2.11 in order to minimize J,

$$u_i = - \sum_{j \in \mathcal{N}_i} \frac{\partial J_{ij}}{\partial \|x_i - x_j\|} \frac{x_i - x_j}{\|x_i - x_j\|} = - \sum_{j \in \mathcal{N}_i} w_{ij}(\|x_j - x_i\|)(x_i - x_j),$$

$$(7.60)$$

for some scalar weight function w_{ij}. As seen in Chapter 2 (Section 2.2), this is nothing but a weighted *consensus* protocol, and it is decentralized insofar as the input, u_i, only depends on Robot i's neighbors.

Setting $h_J(x) = -J(x)$, the construction from the previous section can now be readily applied to the minimization of the cost in Equation 7.59 by formulating the minimum-energy problem,

$$\min_{u, \delta} \|u\|^2 + \delta^2 \quad \text{subject to} \quad \frac{\partial h_J}{\partial x} u \geq -\alpha(h_J(x)) - \delta, \qquad (7.61)$$

where $u = [u_1^T, \ldots, u_N^T]^T \in \mathbb{R}^{Nd}$. Obviously, $\dot{x}_i = u_i$ is drift-free, so solving the optimization problem in Equation 7.61 leads to the accomplishment of the task, by which we mean that a stationary point to the cost, J, has been reached (asymptotically or in finite time), per Corollary 7.1.

But, just because the problem can be solved, it does not follow that it can be solved in a decentralized manner by the individual robots themselves. What we would like is for each robot to solve its own local version of the problem in Equation 7.61, i.e.,

$$\min_{u_i, \delta_i} \|u_i\|^2 + \delta_i^2 \quad \text{subject to} \quad \frac{\partial h_i}{\partial x_i} u_i \geq -\alpha(h_i(x)) - \delta_i, \qquad (7.62)$$

where

$$h_i(x) = - \sum_{j \in \mathcal{N}_i} J_{ij}(\|x_i - x_j\|). \qquad (7.63)$$

To get a handle on when this decentralization is possible, the derivation in Equations 7.46—7.51 ensures that by imposing the global CBF constraint

$$\frac{\partial h_J}{\partial x} u \geq -\alpha(h_J(x)) - \delta, \qquad (7.64)$$

constructed using the whole state vector, x, the cost, J, is decreasing towards a stationary point. We want to show that by imposing only local constraints (i.e., such that Robot i only needs information about its neighbors), the multi-robot system is still able to enforce the global constraint in Equation 7.64 and hence to minimize the cost, J, in a decentralized fashion.

To this end, we first assume that the extended class $\mathcal{K}_\infty$-function, α, is *super-additive* for negative arguments, in the sense that

$$\alpha(z_1 + z_2) \geq \alpha(z_1) + \alpha(z_2), \quad \forall z_1, z_2 < 0. \tag{7.65}$$

Under this assumption, we can sum up the constraints in Equation 7.62 over all robots,

$$\sum_{i=1}^{N} \left(\frac{\partial h_i}{\partial x_i} u_i \right) \geq \sum_{i=1}^{N} (-\alpha(h_i(x)) - \delta_i)$$

$$\geq -\alpha \left(\sum_{i=1}^{N} h_i(x) \right) - \delta = -\alpha \left(h_J(x) \right) - \delta, \tag{7.66}$$

where we used the super-additivity property of α, and we set

$$\delta = \sum_{i=1}^{N} \delta_i. \tag{7.67}$$

Moreover, since the graph, G, which encodes the adjacency relationships between robots, is undirected, we have that

$$\frac{\partial h_i}{\partial x_i} = \frac{\partial h_J}{\partial x_i}. \tag{7.68}$$

Thus, in light of Equation 7.66,

$$\frac{\partial h_J}{\partial x} u \geq -\alpha \left(h_J(x) \right) - \delta. \tag{7.69}$$

Hence, x will converge to a stationary point to J, and decentralization has been achieved. The only price one has to pay for this to be possible is the super-additivity restriction imposed on α.

Circling back to the discussion about finite-time performance, we could select

$$\alpha(z) = cz^\gamma, \tag{7.70}$$

which is a convex function for $z < 0$, and hence it is indeed super-additive for $z < 0$. What we have done is establish the following theorem, which is available in [306] in a slightly different form.

Theorem 7.4 Given the additive, pairwise cost function, J, defined in Equation 7.59. A collection of N robots with single integrator dynamics will minimize J in a decentralized fashion if each robot executes as control input the solution to the optimization problem:

$$\min_{u_i, \delta_i} \|u_i\|^2 + \delta_i^2 \text{ subject to } \frac{\partial h_i}{\partial x_i} u_i \geq -\alpha(h_i(x)) - \delta_i. \qquad (7.71)$$

Here, $h_i(x)$ is given in Equation 7.63, and α is an extended class $\mathcal{K}_\infty$-function that additionally is super-additive for negative arguments. Moreover, if $\alpha(z) = cz^\gamma$, for some $c > 0$, $\gamma \in (0, 1)$, a stationary point to the cost, J, is reached in finite time, with an upper bound on the settling time given by Theorem 7.3.

The structure of the cost function, $J(x)$, even though quite specific, supports a rich set of multi-robot behaviors, as shown in Chapter 2. In the next section, we illustrate this point by considering the formation and coverage control problems in this new light.

7.3.1 Formation and Coverage Control Revisited

In formation control applications, the robots are tasked with assembling a particular shape, typically specified in terms of inter-robot distances. The pertinent, pairwise performance cost, borrowed from [93, 306], is given by

$$J(x) = \frac{1}{2} \sum_{i=1}^{n} \sum_{j \in \mathcal{N}_i} \frac{1}{2} (\|x_i - x_j\| - d_{ij})^2 = \sum_{i=1}^{n} J_i(x), \qquad (7.72)$$

where d_{ij} is the desired distance between Robots i and j, and $J(x) = 0$ corresponds to the robots being perfectly in formation.

The gradient to $J_i(x)$ evaluates to

$$\frac{\partial J_i}{\partial x_i} = \sum_{j \in \mathcal{N}_i} \frac{\|x_i - x_j\| - d_{ij}}{\|x_i - x_j\|} (x_i - x_j), \qquad (7.73)$$

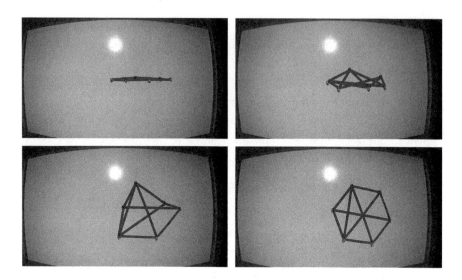

Figure 7.3: A team of robots in the Robotarium testbed [337] executes the formation control strategy obtained when using Equations 7.72 and 7.73 as inputs to the optimization program in Theorem 7.4. The relevant edges, whose distances are to be driven to desired values, and subsequently maintained, are projected onto the testbed.

which has a rather direct interpretation when employed in a steepest descent flow. If the distance between Robots i and j is smaller than d_{ij}, then the weight

$$w_{ij}(\|x_i - x_j\|) = \frac{\|x_i - x_j\| - d_{ij}}{\|x_i - x_j\|} \tag{7.74}$$

is negative and the robots repel each other. Conversely, if the two robots are farther than d_{ij} apart, the positive weight will result in an attraction between the robots.[13]

The effect of applying the constraint-based formalism, as detailed in Theorem 7.4, to this particular choice of performance cost, is shown in Figure 7.3. There, six robots are tasked with assembling a hexagonal shape, as specified through a rich enough subset of all possible inter-robot distances to render the formation *infinitesimally rigid* [15, 134], i.e., the pairwise distance

---

[13]The special case when $d_{ij} = 0$, $\forall i, j$, corresponds to a negative gradient flow that coincides exactly with the previously discussed consensus protocol [197, 276].

specifications are sufficient to completely determine the shape, up to rotations and translations.

Formation control, as generated by pairwise inter-robot distances, fits the formulation in Equation 7.63 extraordinarily well. However, upon further inspection, Theorem 7.4 only requires that the cost be given as a sum over individual (and locally evaluable) performance costs. The pairwise structure is strictly speaking not necessary, as observed in [306]. To illustrate this point, we leave the formation control domain, and move on to the question of how to achieve effective multi-robot area coverage through constraints rather than costs.

Coverage control, as we have seen, is concerned with the problem of having multiple robots cover an area, D. Given a coverage performance cost, the robots should spread out over the domain in order to minimize the cost. As shown in [245], each robot could, for example, be responsible only for the subset of the domain that is defined by its Voronoi cell, $V_i = \{p \in D \mid \|p - x_i\| \le \|p - x_j\| \ \forall i \ne j\}$.

A performance cost for the domain coverage problem was provided in [306],

$$J(x) = \sum_{i=1}^{N} \frac{1}{2} \|x_i - \rho_i(x)\|^2 = \sum_{i=1}^{n} J_i(x), \qquad (7.75)$$

where $\rho_i(x)$ is the centroid of Robot i's Voronoi cell.[14] Taking the derivative of J_i yields

$$\frac{\partial J_i}{\partial x_i} = (x_i - \rho_i(x))^T \left(I - \frac{\partial \rho_i(x)}{\partial x_i} \right), \qquad (7.76)$$

where I is the identity matrix.

The derivative in Equation 7.76 can now be inserted directly into Equation 7.71 in Theorem 7.4, and the application of the constraint-based, robot ecology machinery to the coverage control problem is depicted in Figure 7.4. As can be seen (and as expected), the six robots converge to a configuration where each Voronoi cell has the the same size and shape and, as such, optimal coverage is achieved.

---

[14]It is worth noting that this performance cost is a reformulation of the locational cost introduced in [95], and provided in Equation 2.19, in that the two costs have the same critical points.

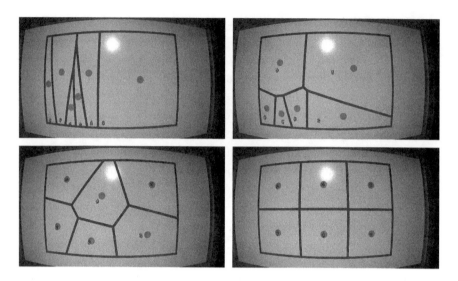

Figure 7.4: Six robots in the Robotarium [337] cover an area using Equations 7.75 and 7.76 as arguments to the constrained optimization problem in Theorem 7.4. The Voronoi cells of the robots are projected onto the testbed, together with their centroids, depicted as gray circles.

7.3.2   Sequential Composition of Behaviors

To see how the finite-time version of the goal-driven barrier function formalism in Theorem 7.4 can come in handy, we briefly consider the problem of stringing together sequences of multi-robot behaviors. The underpinning idea is that by decomposing high-level specifications into sequences of dedicated multi-robot behaviors, the design complexity is potentially reduced quite significantly. This decomposition idea is particularly advantageous as one can tap into a plethora of readily available multi-robot behaviors with provable performance guarantees, such as flocking [316], rendezvous [317], formation control [135], and so on.

The main challenge associated with producing sequential compositions of multi-robot behaviors is, as we have seen, that most behaviors rely strongly on particular, underlying network topologies in order to function properly, e.g., [276]. Therefore, in order to meet the high-level specifications through sequences of behaviors, not only should each of the individual behaviors be correctly chosen, but the network assumptions should also be satisfied—both at the starting time of the behavior, and throughout its execution.[15]

---

[15]Decompositions of complex multi-robot missions have been explored using a number of other formalisms, such as Petri nets [212], finite state automata [260], stochastic

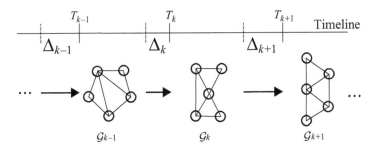

Figure 7.5: A segment of an MDL sequence of multi-robot behaviors for a team of five robots. Behavior k starts at time τ_k, and requires a network topology, encoded by the graph $\mathcal{G}_k$, in order to function properly. To prepare for the next behavior, the corresponding finite-time barrier constraint has the preparation time, Δ_{k+1}, at its disposal, during which time Behavior k is modified through the inclusion of an additional network topology constraint, so as to establish the required graph structure, $\mathcal{G}_{k+1}$, by time τ_{k+1}.

To illustrate the general setup, consider the construction introduced in [235], based on the concept of a *motion description language* (MDL) [262] to specify the requisite sequence of behaviors, namely,

$$\sigma = (\mathcal{U}_1, \mathcal{G}_1, \tau_1), \cdots, (\mathcal{U}_M, \mathcal{G}_M, \tau_M), \tag{7.77}$$

where each multi-robot behavior is abstractly represented by the tuple $(\mathcal{U}_k, \mathcal{G}_k, \tau_k)$, $k \in [M]$.

In the MDL string in Equation 7.77, $\mathcal{U}_k$, $\mathcal{G}_k$, and τ_k represent the multi-robot controller, required interaction graph structure, and starting time of Behavior k, respectively. The required graph structure, $\mathcal{G}_k$, is additionally given by (V, E_k), and in order to ensure that Behavior k is executed correctly, the required graph, $\mathcal{G}_k$, should remain a spanning subgraph to the actual interaction graph, $\mathcal{G}(t) = (V, E(t))$, throughout the execution of the behavior, i.e., $(i, j) \in E(t), \forall (i, j) \in E_k, t \in [\tau_k, \tau_{k+1})$.

An example of a such an MDL sequence is shown in Figure 7.5. In that figure, Behavior k is executed without consideration of the next behavior until time $\tau_{k+1} - \Delta_{k+1}$, where τ_{k+1} is the time when the system is supposed to switch to Behavior $k + 1$. That leaves a time window of length Δ_{k+1} for the system to prepare for the behavior transition by assembling the

---

processes [291], temporal logic specifications [213, 363], embedded graph grammars [211, 382], and graph process specifications [414].

required topology, encoded through the edge set in $\mathscr{G}_{k+1}$. Using the finite-time formulation in Theorem 7.4 (including equating Δ_{k+1} with the settling time in Theorem 7.3), this approach was leveraged in [235] to formally compose together sequences of multi-robot behaviors with provable performance guarantees in order to support a number of complex multi-robot specifications.

Based on [235], a simulated team of six robots is tasked with an environmental exploration task in Figure 7.6. Additionally, the same sequencing framework is experimentally deployed on robots in the Robotarium [337] in Figure 7.7. In the experiment, the robot team executes a sequence of multi-robot behaviors consisting of a cyclic pursuit behavior, a lattice formation, a line formation, and a square formation. Figure 7.7 contains a sequence of snapshots taken during the experiment, where the edges prescribed by each behavior do indeed remain in place throughout the execution of the behavior, and the additional edges required by the following behavior are established during the requisite preparation period.

7.4   Putting It All Together

We have now arrived at the point where, what began with a few anecdotal observations about Mars rovers in Chapter 1, a mathematically precise robot ecology formalism has been established. The beginning of the story focused on the tight coupling between robots and their environments, which is the reason why the ecological notes were struck in the first place. Additionally, it was discovered that power-management was highly integral to long-duration autonomy, and that surviving was a prerequisite to thriving.

By leveraging the spirit (if not the formalism) of behavior-based robotics (Chapter 2), and solidifying the ecological connections (Chapter 3), constraints emerged as the key drivers of behavioral expressivity when deploying robots over long time-scales. In the second part of the book, we came to the realization that control barrier functions provide the right tool for such a shift towards constraint-based control design. As such, *robot ecology* is no longer shrouded in mystery, and in this concluding section of the chapter, we gather together all the pieces as well as foreshadow two canonical instantiations of the robot ecology idea.

7.4.1   A Purposeful Yet Safe Expenditure of Energy

If at all possible, the only cost to be minimized should be the energy consumed. All other considerations should be relegated to constraints. Foremost

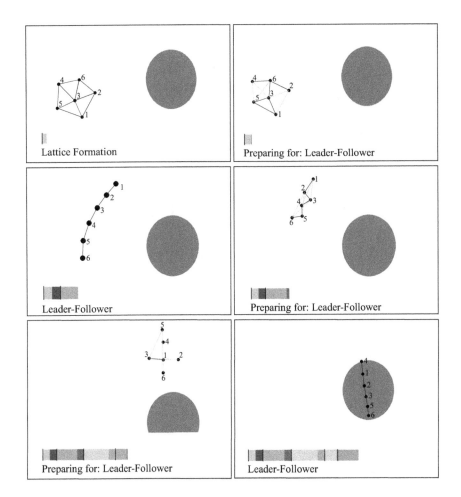

Figure 7.6: A sequence of multi-robot behaviors designed for an environmental exploration task. The team of six robots is tasked with investigating a region of interest, displayed as an ellipse. The robots initially gather together in a lattice formation, after which a sequence of different "leader-follower" formations are used to move the robots close to the region of interest, to prepare the team for entering the region. The mission is concluded with a final leader-follower behavior, where the leader-robot leads the team inside the region. Source: Fig. 3 in Li et al. (2018).

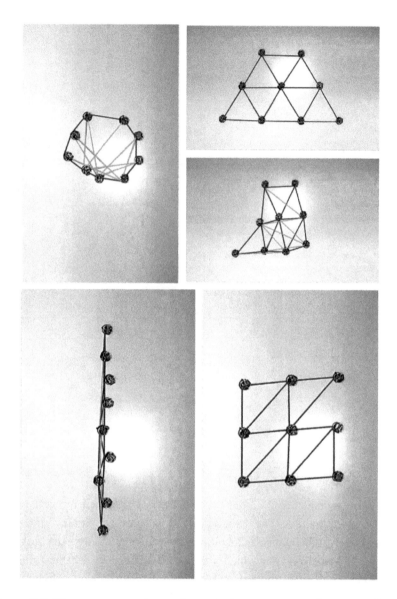

Figure 7.7: Shown are experimental results for a team of nine robots executing a sequence of multi-robot behaviors in the Robotarium [337], with the edges superimposed on the robot arena by an overhead projector. The edges prescribed by each behavior are shown together with the additional edges required by the next behavior (as established during the preparation time). Depicted is a cyclic pursuit behavior (Top Left), followed by a sequence of geometric formations (lattice, line, and square).

of those constraints is survivability. Although this constraint may come in different sizes and shapes, as per Section 7.1.2, some version of it must always be present to ensure robot longevity.

In addition to survivability, exogenous tasks may be added to the robots' list of responsibilities. Ultimately, the robots are supposed to do something useful. At least every now and then. And, we have now seen that it is possible to encode such tasks as constraints as well, as long as the task performance can be evaluated through an instantaneous performance cost.

We here gather these observations, as well as refer back to the five Design Principles for Robot Ecology used to start off the discussion in the chapter. The result from this exercise is a formal framework for the *purposeful yet safe expenditure of energy for robots deployed in natural environments over long time-scales!* In other words, the challenge laid down in Chapter 1 has been met.

Robot Ecology

Central to any robot deployed in an environment over long time-scales is survivability, as per Principles *P1*, *P3*, and *P4*. The general survivability constraint, as seen in Equation 7.24, is given by the composite barrier function

$$h_{survive}(x) = h_{resources}(x) \wedge h_{interact}(x) \wedge h_{homeostasis}(x), \qquad (7.78)$$

where the particulars of the constituent constraints may vary as functions of factors such as environmental conditions, robot types, and intended missions.

In Between Missions
Principles *P2* and *P5* dictate that in between missions, the robots should simply be present in the environment, surviving, and be prepared to be recruited.
The corresponding constrained optimization problem is a pure, minimum-energy problem,

$$\min_{u} \ \|u\|^2 \qquad\qquad (7.79)$$

subject to

$$S_{h_{survive}}[x; u] \geq 0. \qquad\qquad (7.80)$$

Missions as Costs

The utility of the robot ecology framework is ultimately derived from its ability to support robots that actually do something useful, i.e., for them to be recruitable to participate in a number of tasks and missions. If the performance can be encoded through the instantaneous task cost, $J_{task}(x) \geq 0$, with the corresponding barrier function being $h_{task}(x) = -J_{task}(x)$, the minimum-energy structure can be kept largely intact also under these scenarios. What the robots should do is solve

$$\min_{u,\delta} \; \|u\|^2 + \delta^2 \tag{7.81}$$

subject to

$$S_{h_{survive}}[x; u] \geq 0 \quad \text{and} \quad S_{h_{task}}[x; u] \geq -\delta. \tag{7.82}$$

Missions as Nominal Inputs

One can envision situations where the tasks cannot be readily framed through task costs. For example, if a human operator takes direct control over a robot, the cost in Equation 7.43 is not directly applicable. Under such conditions, one can instead let the mission be interpreted in terms of a nominal control input, u_{nom}. As a result, the minimum-energy cost is no longer fully supportable, and the mismatch between actual and nominal control inputs should instead be minimized,

$$\min_{u} \; \|u - u_{nom}\|^2 \tag{7.83}$$

subject to

$$S_{h_{survive}}[x; u] \geq 0. \tag{7.84}$$

An example of the second of these three cases—Missions as Costs—is shown in Figure 7.8. In the example, a team of robots is tasked with covering an area, using the coverage cost in Equation 7.75. At the same time, they are to avoid collisions, return to charging stations when low on battery, and charge until the battery levels are sufficiently high. In other words, what the robots are executing is the controller resulting from the following constrained

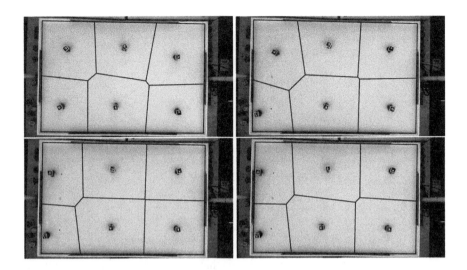

Figure 7.8: A sequence of salient frames from a persistent monitoring scenario framed as a "Missions as Costs" coverage problem is shown. Six robots are deployed in the Robotarium [337]. Projected onto the Robotarium arena are the Voronoi cells associated with each robot. In the progression, the top left robot is moving towards the charging strip at the left edge of the arena in order to recharge.

optimization program:[16]

$$\min_{u,\delta} \|u\|^2 + \delta^2$$
$$\text{subject to } S_{h_{survive}}[x; u] \geq 0 \text{ and } S_{h_{task}}[x; u] \geq -\delta, \tag{7.85}$$

where

$$h_{task}(x) = -\frac{1}{2} \sum_{i=1}^{N} \|p_i - \rho_i(p)\|^2, \tag{7.86}$$

with $p_i \in \mathbb{R}^2$ being the position of Robot i, $i \in [N]$, and $\rho_i(p)$ is the centroid of its Voronoi cell. Additionally,

$$h_{survive}(x) = \min\{h_{resources}(x), h_{interact}(x), h_{homeostasis}(x)\}, \tag{7.87}$$

---

[16]The robots in Figure 7.8 are using the distributed versions of the task and collision constraints, as per Equation 7.63 and Equation 7.3, respectively. We have opted to not spell this out explicitly, for the sake of notational clarity and convenience.

where the constituent constraints are given by

$$h_{resources}(x) = \min\{h_{obstacle}(x), h_{anchor}(x)\}$$

$$h_{obstacle}^{ind}(x_i) = \min_{j \in [O]} \{\|p_i - o_j\|^2 - D_o^2\}$$

$$h_{anchor}^{ind}(x_i) = (E_i - E_{min}) - \mathcal{E}(dist(p_i)) \geq 0$$

$$h_{interact}(x) = h_{separation}(x) \qquad\qquad (7.88)$$

$$h_{separation}^{pair}(x_i, x_j) = \|p_i - p_j\|^2 - D_s^2$$

$$h_{homeostasis}(x) = h_{energy}(x)$$

$$h_{energy}^{ind}(x_i) = (E_{max} - E_i)(E_i - E_{min}).$$

Here, E_i is Robot i's energy level that must remain in the interval $[E_{min}, E_{max}]$, $[O]$ is the index set of obstacles in the environment, $\mathcal{E}(dist(p_i))$ is the energy required to travel to the closest charging station, and D_s and D_o are safety separation distances, as per the discussion in Section 7.1.

In Figure 7.8, six robots in the Robotarium [337] are instrumented with two extended prongs at different heights that, when charging, connect to two aluminum strips embedded in the arena walls. One of the metal strips supplies a 5V input voltage to the robots, while the other serves as ground, which allows the robots to drive up to the charging stations and recharge without human intervention. The distance to the closest charging location, $dist(p_i)$, in Equation 7.88, is thus the distance to the point on the charging strip that is closest to the robot. To explicitly call out the fact that robot longevity is achieved in terms of never running out of battery, the measured battery voltage data are shown in Figure 7.9.

7.4.2 The End Game

With that, the technical developments in the book have concluded. Rather than introduce new mathematical tools, the final two chapters describe canonical instantiations of the robot ecology idea. To call out particularly key, salient features, these two instantiations are selected to highlight issues pertaining to persistent environmental monitoring and autonomy-on-demand.

In Chapter 8 a long-duration environmental monitoring scenario is being considered. In particular, the SlothBot [303], which is a slow and energy-efficient, wire-traversing robot, is operating among the treetops for sustained

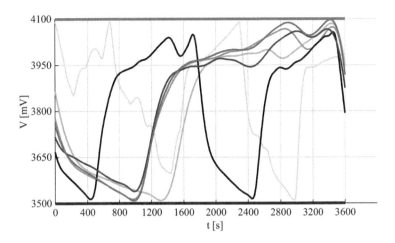

Figure 7.9: Measured battery voltage data for each of the six robots in Figure 7.8. The top and bottom horizontal lines depict the values of $E_{\min}$ and $E_{\max}$ in h_{energy} and h_{anchor} in Equation 7.88, respectively. As expected, the voltages always stay inside the prescribed region.

Figure 7.10: The SlothBot (Left) and Robotarium (Right) constitute canonical instantiations of the robot ecology concept, designed for long-duration environmental monitoring and autonomy-on-demand, respectively.

periods of time, as seen in Figure 7.10 (Left). Most of that time is spent doing mostly nothing other than hanging on wires in the tree canopies, taking measurements, and collecting environmental data. But, every now and then, the SlothBot leaves the shaded canopies to go and "sunbathe" and recharge the batteries, thereby embodying the long-duration autonomy concept.

Chapter 9, in turn, focuses on the autonomy-on-demand ideas that have been hinted at repeatedly throughout the book. Exemplified by the Robotarium testbed [337], which is an always-on, remotely accessible swarm robotics lab, a large team of robots are sitting idle, waiting to be recruited to perform whatever tasks remote users may require of them. The Robotarium, shown in Figure 7.10 (Right), has been in (more or less) continuous operation over multiple years, and has supported thousands of remote experiments submitted by users from all over the world [431]. As such, it constitutes a crisp, canonical instantiation of the autonomy-on-demand approach to robot ecology.

8 Environmental Monitoring

Having established a general robot ecology framework where constraint-driven control enables robots to be present and survive in natural environments over long periods of time, it is time to return to one of the more pertinent, underlying application-drivers in greater detail. In this chapter, we consider *environmental monitoring*, where robots are to be deployed in an area, tasked with doing nothing beyond collecting data that can be of use to fields such as precision agriculture, conservation and environmental research, or, more broadly, climate-related studies. To that end, this chapter is initially focused on how to tailor the robot ecology framework to questions pertaining to persistent environmental monitoring, followed by a treatment of a particular, energy-efficient environmental monitoring robot, namely the *SlothBot*. The SlothBot is a solar-powered, wire-traversing robot, suspended in the tree canopies. As such, it will serve as a canonical instantiation of the broader persistent environmental monitoring theme espoused in this chapter.

As the primary focus in this chapter is on issues surrounding persistently present robots rather than on multi-robot interactions, we here consider the case of a single robot (the SlothBot). As a consequence, the environmental monitoring task will *not* proceed according to the persistified coverage control program introduced in Chapter 5 for multi-robot teams. Instead, the long-duration autonomy deployment will adhere to the broader robot ecology formalism from the previous chapter in that the robot will, most of the time, solve the constrained, minimum-energy problem,

$$
\begin{aligned}
&\min_u \ \|u\|^2 \\
&\text{subject to } \ S_{h_{survive}}[x; u] \geq 0.
\end{aligned}
\tag{8.1}
$$

As before, the survivability constraint is given by

$$
h_{survive}(x) = h_{homeostasis}(x) \wedge h_{resources}(x) \wedge h_{interact}(x),
\tag{8.2}
$$

with the particular, constituent constraints being

$$
\begin{aligned}
&h_{homeostasis}(x) = h_{energy}(x), \quad h_{resources}(x) = h_{obstacle}(x) \wedge h_{anchor}(x), \\
&h_{interact}(x) = h_{identity}(x),
\end{aligned}
\tag{8.3}
$$

as per Equation 7.24. Since the constraints in Equation 8.3 are discussed in technical detail in Chapter 7, we simply refer back to those discussions when needed, which allows us to instead focus on how to make such an abstract formulation come "alive" and be realized on the SlothBot in a practical, environmental monitoring scenario.

8.1 Monitoring in Natural Environments

Ecology, as a scientific discipline, touches on the many intricate, intertwined, and complex interconnections formed between different organisms, and between organisms and their environments, e.g., [308, 357]. These interconnections are formulated in terms such as consumer-resources, predator-prey, mutualism, or competitive relationships, as discussed in Chapter 3. But, actually characterizing such intricate relationships is only possible if sufficient data is available, i.e., if enough meaningful observations can be made about what really goes on in an ecosystem. To that end, field ecologists spend significant effort on the collection of relevant data; some of the practical mechanisms by which this can be achieved are highlighted in the next section. Such a practical detour is needed as it will help inform the design of environmental monitoring robots for the purpose of ecological research in general, and conservation biology among the treetops in particular. It will also allow the robot ecology framework, that has benefited significantly from the field of ecology, to provide some direct benefits back to practicing ecologists.

Monitoring, viewed as a conceptual activity, involves the gathering of data in some environment of interest. But, it becomes meaningful only in some broader context in which the data is actionable. *Conservation biology* is a prime example of such a context, as it is a discipline focused on the management of nature and of Earth's biodiversity [341]. The ultimate aim with this field of study is to protect species, habitats, and ecosystems. And to do so successfully requires data.

Conservation biology is not only a suitable motivation for the robotic environmental monitoring tasks considered in this chapter, it is also an important endeavor in its own right. Ecosystems all across the planet are in rapid decline, which means that conservation biology is associated with a strong sense of urgency. In fact, [430] refers to conservation biology as a "discipline with a deadline." Along the same lines, in [215], it is suggested that 50% of all species will disappear towards the middle of the century, which has potentially very serious consequences for the stability of the planet [169, 403]. For example, the destruction of forest ecosystems alone account for over 10% of all greenhouse gas emissions caused by humans [164]. As trees store a significant amount of carbon—thereby keeping it out of the atmosphere—conserving and protecting forest ecosystems has a real and positive impact on the climate. Forest ecosystems are highly complex, however. This means that the protection needs to extend to their overall, supported biodiversity, beyond "just" the trees themselves.

For these reasons, conservation biology is a highly worthy motivation for the monitoring activities considered in this chapter, while forests constitute an important and suitable habitat for the monitoring robots. To that end, we will particularly focus on robots present in tree canopies for long periods of time. Before we can get to the robots themselves, however, we need to investigate what it is that they should be monitoring in the first place.

8.1.1 Biodiversity

One crucial manifestation of the overall health and robustness of an ecosystem is its *biodiversity*. This concept is attempting to encompass the variety and variability of life on Earth, which means that it has to be measured at a number of different scales, such as at genetic, species, and ecosystem scales, [155, 184, 289], with biodiversity somehow being the totality of genes, species, or ecosystems in a given region [108]. For example, if one were to count the number of species (or some other relevant taxa) in a habitat, and then repeat this process for all individual habitats in a particular region, then the so-called α-diversity would be the mean species diversity per habitat, the γ-diversity would be the total species diversity over the entire region, and the β-diversity is the ratio between regional and local species diversity, i.e., $\beta = \gamma/\alpha$ [412].

But, such relationships between different biodiversity notions only make sense if the arguments themselves are quantifiable, i.e., we need to understand what we mean when we say things like "mean species diversity." This modest request is unfortunately not as easy to accommodate as one might expect. In fact, a number of different measures of "diversity" have been proposed, such as Species Richness, which is simply the number of different species. This is arguably not the most interesting of measures, as an ecosystem consisting of 999 members of Species A and 1 member of Species B would be considered as diverse as one consisting of 500 of each species, as the Species Richness is 2 in both cases. To address this issue, a handful of other measures have been proposed, such as the Shannon Index (i.e., entropy), the Simpson Index, and the Berger-Parker Index; all of which quantify diversity based on (different twists on) the percentage of the total population belonging to each species, e.g., [387]. For example, if p_i is the probability that a randomly selected individual belongs to Species i, then the Shannon Index is

$$\mathcal{H} = -\sum_i p_i \ln p_i. \tag{8.4}$$

This means that if we return to the two-species thought-experiment, $p_A = 0.999$ and $p_B = 0.001$ gives $\mathcal{H} \approx 0.0079$, while $p_A = p_B = 0.5$ gives $\mathcal{H} \approx 0.69$. The latter situation is significantly more diverse, as expected from an entropy-based biodiversity measure.

Pure probability-based measures do suffer from their own set of issues that may or may not be problematic, depending on the context. These issues stem from the fact that they do not take into account how *different* the species are from one another. Returning to the situation where $p_A = p_B = 0.5$. If Species A is a dark-blue butterfly and Species B is a light-blue butterfly, then this situation could be argued to be less diverse than a situation where Species A is a butterfly while Species B is a lion.

To explicitly call out the differences between species present in an area, Rao's Quadratic Entropy [349] incorporates these differences in the diversity measure. Given a distance measure defined over the set of species, $dist(i, j) \in \mathbb{R}_{\geq 0}$, the quadratic entropy is given by the expected distance-squared between any two randomly chosen members of the population, i.e.,

$$\mathcal{Q} = \sum_{i,j} p_i p_j \text{dist}(i,j)^2. \tag{8.5}$$

Assuming that such a distance function can be meaningfully defined and that, for example, the "distance" between the two butterfly species is 1, while the "distance" between a butterfly and a lion is 10, then the former case has $\mathcal{Q} = 0.25$, while the latter has $\mathcal{Q} = 25$, i.e., it is significantly more diverse.[1]

For the environmental monitoring tasks under consideration in this chapter, what perhaps matters more than how biodiversity is ultimately defined is how the data is obtained in the first place. Once one has access to the relevant data, one can proceed to argue over which measure to use to quantify diversity.

Biodiversity fieldwork is inevitably concerned with some form of cataloging the number and variety of organisms found in a specific region. The most basic question associated with such an activity is how many (different kinds of) organisms there are in a given area. Currently, there are close to 10 million known species identified on Earth, but most estimates point to an actual number closer to 100 millions (and beyond), e.g., [246, 283]. This is a remarkably large and uncertain number, and the uncertainty stems from the difficulty associated with conducting a reliable species census. Access is one problem. But so is making sure that the observations are persistent and comprehensive enough in order to ensure that as many species as possible are encountered in a given area.

There are a number of practical approaches taken when it comes to measuring biodiversity data,[2] with the most common ones including canopy fogging, quadrat sampling, transect sampling, netting, and trapping [247, 359, 394]. The particular method employed is highly dependent on its suitability for a particular environment or target species; clearly counting insects in a rainforest is a quite different undertaking than counting sea urchins on a coral reef. In fact, canopy fogging is particularly effective when it comes to gathering information about insects. Basically, a low (preferably non-harmful)

---

[1]Within the multi-robot community, similar notions have popped up when it comes to measuring how heterogeneous a group of robots is, where notions such as the number of robot types (species) and the differences between robots also play a key role, e.g., [35, 342, 415].

[2]We here take a bit of liberty with the term "biodiversity" to let it encompass quantities associated with other lines of inquiry as well, such as organism ecology, where a complete census may neither be required nor desired, but where individual organisms need to be cataloged and tracked, nonetheless [182].

dose of insecticide is sprayed in the tree canopy. This knocks out the insects that subsequently fall towards the ground to land on a funnel-shaped screen, thereby ensuring that a majority of the insects are collected and accounted for.

For plants or animals that are ground-based and relatively slow, quadrat sampling is one method of choice. A quadrat is simply a square area—typically ranging from $1m^2$ to $20m^2$. Practically, such a square is staked off, followed by a careful inventory of the area in order to catalog all inhabitants in the quadrat. This time-consuming and painstaking process is typically done manually, and it is repeated in multiple locations in the same general area to get as close as possible to an accurate biodiversity count [359, 394].

A variation on this theme is to follow a line through an area, rather than going over a square. This is known as transect sampling, and the operation involves rolling out a line (like a rope or tape measure), marked at regular intervals, and then, at every interval, recording the number of species/organisms encountered. Such a strategy is particularly effective in certain aquatic ecosystems, such as coral reefs. For speedier organisms, neither of these mechanisms are adequate. For example, birds and bats in terrestrial ecosystems will clearly not stay still long enough to be counted in this manner. Instead, netting or trapping techniques are used, where animals are captured for the purpose of being counted, identified, cataloged, and tracked.

When cataloging terrestrial mammals, the traps are typically arranged in an "array," i.e., in a particular geometric configuration, such as a grid or a web [359]. Captured animals are tagged and released, which provides field ecologists with information such as the age of the captured individuals, which can serve as a strong indicator of a growing or declining population. Additionally, by tracking the animals over time, movement patterns, mortality rates, or survival/reproductive probabilities can be deduced. But, as netting or trapping strategies are typically non-comprehensive in that not all animals in an area are reached, a single such capture campaign does not provide an answer to the question of how many animals there are—only how many were captured.[3] Under the assumption that the capture probabilities are independent, the way this conundrum can be overcome is through a so-called capture-recapture strategy. Assume that M animals are captured during the first phase. These animals are tagged and, subsequently, released. Then, during the recapture phase, N

---

[3]This difficulty is compounded by the ecology adage that *rare species are common and common species are rare*, i.e., the coverage achieved through trapping may be spotty at best.

animals are captured, of which $M' \leq M$ were captured (and tagged) also during the first phase. The fraction $p = M'/M$ thus provides us with a (very) rough estimate of the fraction of individuals that are captured during each attempt, which, in turn, translates to an estimate of the total number of animals in the area, given by N/p, or NM/M', e.g., [387].[4]

These approaches provide different practical avenues for counting organisms. But, such cardinality-based quantities alone clearly do not tell the whole story of what goes on in a given area. Instead, they need to be coupled to other environmental factors, such as climate and weather conditions, in order to support richer, so-called *ecological niche models*, e.g., [130, 206]. Such models, if constructed correctly, possess the types of predictive powers that allow conservation biologists to go from reactive to proactive policies for the purpose of protecting Earth's ecosystems.

8.1.2   Microclimates and Ecological Niche Models

If, on the one hand, an effective environmental monitoring robot needs to be able to detect, count, and catalog distinct organisms for the purpose of quantifying biodiversity or other related organism-centric notions,[5] it also needs to be able to measure factors relevant to the local environment and microclimate surrounding the area in which the organisms are present. Defined as the climate of a restricted, well-delineated, and (relatively) small area [360], microclimate data can help support the development of useful and predictive conservation models.

As there is no unique notion of what constitutes a "small' area, the range over which a microclimate is defined is linked to the types of questions one is interested in—from mountains and deserts all the way to the insides of carnivorous pitcher plants [200]. What this means is that the scale of the habitat of the organism under consideration needs to line up with the scale of the measurements, may they be for microclimate or other ecological purposes. What might be highly relevant, or even existential, to a pollinating insect, could simply be noise to an elephant. This idea of pursuing measurements at the appropriate scale relative to the conservation goals resonates well with

---

[4]More sophisticated models can be derived, and more capture phases can be introduced. But the gist behind the strategies are essentially the same, e.g., [255].

[5]These related inquiries of relevance to studies of organisms, populations, and communities could include questions such as: Where are the animals? What are they doing? What are they eating?

one of the central tenets of robot ecology in that not only should the tight robot-environment connections be embraced, if the robot is pursuing environmental monitoring tasks, the observation-environment connections must be considered as well.

Regardless of what the scope is and what type of microclimate is being probed, the typical set of measurements include temperature (using a thermometer), atmospheric pressure (barometer), humidity (hygrometer), precipitation (rain gauge), solar radiation (pyranometer), air quality, and wind speed (anemometer) and direction (wind sock/wind vane). Climate measurements at sea nominally include other quantities as well, such as wave height and frequency, while observations taken in forest ecosystems try to capture additional factors, such as leaf wetness and soil moisture and temperature [182]. A fairly standard set of instruments is associated with these different types of measurements, and the relevant subset must be present on an environmental monitoring robot.

The microclimate and other environmental conditions in an area are only partially characterized and measured through these mentioned, *abiotic* factors, however. The conditions are likewise heavily influenced by features in the landscape itself, such as the topography, i.e., the overall shape of the area, including hill tops or valleys, the soil composition, e.g., the soil's ability to absorb and retain water, the general presence of water (such as rivers or lakes), the type and density of vegetation, and the presence of artificial structures, e.g., [151]. In fact, since the landscape "geometry" plays such an important role for determining the suitability of a habitat to a particular species, this type of information is important to gather also when one is looking at broader ecological or conservation biology questions, beyond just microclimates.

As biodiversity and wildlife demography ultimately boil down to quantifiable measures—How many species are there?—and microclimate models capture general environmental conditions, they do not necessarily, by themselves, provide the kind of rich models needed to untangle and describe the many complex interconnections that make up an ecosystem. In fact, *if one thinks of the organisms as the actors, and the environment as the stage, ecology is what happens when the performance starts, i.e., when the actors take the stage.*

Ecological niche models approach this level of texture in that they constitute a class of models that combine the observations made under the biodiversity umbrella, like the occurrence of individuals of certain species, with environmental and microclimate data. This combination of models can further be used to generate so-called correlative models, i.e., models that interpolate the abundance (or lack thereof), reproductive success, or mortality rates of a particular species across environmental variables [379]. The result is a model of the environmental conditions that meet a species' ecological requirements, as well as provides predictions of the relative suitability of a given habitat. If one wants to answer questions like, "Why is the black bear population declining on a particular peninsula?" then past, present, and future connections must be considered between the bear population, the environment, and other members of the community.[6]

An ecological niche model's efficacy when it comes to predicting the distribution of species across varying environmental conditions depends on the prevalence of high-quality data. As seen, the environmental data are frequently climate-related data, such as temperature and precipitation, but can also include other factors such as soil type, water depth, and land cover. The connection between data and conservation biology is derived from the predictive power of the model in that one can (at least in theory) predict potential future distributions of invasive species or the expected success when reintroducing vulnerable species to an area [130], as pictorially illustrated in Figure 8.1.

8.1.3 Under the Tree Canopies

We have now seen a number of more or less systematic processes for collecting data relevant to conservation biology. But, in a number of environments, it is simply not feasible to stake out a square in order to completely catalog all life encountered in that quadrat, or to knock out the animals with insecticides. Instead, direct, "manual" observations will have to suffice. This is certainly true for tree canopies, where it is hard to get access and to perform long-term observations. This relative lack of ready access is in stark contrast to the outsized importance that forest canopies play as the home to a sizable share of Earth's species; tree canopies are where solar energy is turned into carbohydrates that, in turn, fuel entire ecosystems [191]. The tree canopies are thought to house at least 50% of the terrestrial biodiversity [429]. In fact, life among the treetops is advantageous in that it is relatively protected, and there is plenty of sunshine and food, such as fruits and flowers.

---

[6]For these correlative models to serve their intended purposes, one must, of course, have access to sufficient observations of the presence of species across environments, as well as absences, i.e., environments where the species cannot be found.

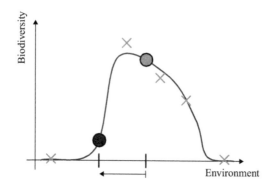

Figure 8.1: Depiction of an ecological niche model. Environmental variables are given on the x-axis, while the abundance of a particular species (or some other relevant biodiversity data) is represented on the y-axis. The Xs are observations; the curve is the correlative, ecological niche model; and the right circle corresponds to the current conditions in a given area. The left circle is a prediction associated with a possible, environmental change. As seen, this particular disruption is not expected to be beneficial to the species under consideration.

In addition, treetops are profoundly important to global climate regulation as well as to the forest microclimates. A tall forest canopy can significantly cool the ground underneath during the day, while at night, the trees absorb significant radiation from the ground, thereby maintaining warmer ground temperatures than what would be the case in the absence of trees. Additionally, forests tend to support low-pressure systems, which, in turn, increases precipitation—it rains more in the forest [191].

It is, at least partially, for these reasons that the forest canopy[7] is sometimes referred to as the "eighth continent of the planet" [250]. And, climate change and biodiversity conservation are but two areas of research that would benefit from access to more data from the treetops. Although some data can be obtained from the ground, the real action in terms of canopy-atmosphere and canopy-forest floor interactions, not to mention an effective cataloging of the forest inhabitants, typically requires a presence within the canopy itself [250].

---

[7]Forest ecosystems are divided into two subsystems, namely the forest floor and the canopy, which, in turn, is defined as the foliated part of the forest, i.e., where the leaves are.

But how does one actually measure what goes on up in the trees? The history of canopy access is very much a technology story. Going from steel towers in Uganda and ladder structures in the Himalayas in the 1960s, to the introduction of single-rope techniques in Australia and Costa Rica in the 1980s, has led to what [250] refers to as the "golden age of canopy access," where the mid-canopy can be reached and observed in a relatively straightforward manner. However, the upper reaches of the canopy is still somewhat elusive, and the range of solutions is quite rich and varied, with data being collected using elaborate ladder structures, e.g., [19], scaffolds or canopy walkways, e.g., [249, 272], construction cranes [327], and even hot air balloons [171]. These techniques have all provided new and useful insights. But, they require a human presence. And, having a persistent, robotic monitoring presence under the treetops would be beneficial. The trees' outsized importance to ecosystems and to the global climate clearly illustrates the need for a good understanding of what actually happens on Earth's "eighth continent," which is where the SlothBot enters the picture.

8.2   Wire-Traversing Robots

We have now shown why the tree canopies constitute a possible sweet spot for an environmental monitoring robot. But, locomoting in such precarious locations is not a trivial undertaking. Ground robots are immediately out of consideration, as they would have a hard time penetrating the tree branches and dense foliage with their sensors, meaning that the observational coverage would be spotty at best. Additionally, forest floors can be quite hazardous, filled with branches, mud, and other obstructions, and moving the robots off the ground would be advantageous also from this long-duration, survival perspective. Aerial robots would certainly solve the ground problem. But they suffer from other issues that make them less than ideal for this particular target application domain. Rotorcraft could potentially hover under the treetops, but this would still be a navigationally tricky proposition. They are, moreover, rather costly from an energy-consumption vantage point, meaning that they would have to fly off to recharge/refuel too frequently for the intended deployment time-spans, which are given in weeks, months, or even years. Fixed-wing or dirigible aerial robots might be energetically suitable, but their maneuverability is simply not up to the task. The former cannot hover, while the latter cannot nimbly follow the wildlife trails, which are two capabilities that are required if one were to effectively monitor life among the trees.

So if the robots cannot be on the ground, nor in the air, where should they be? Well, they could climb. This is what arboreal animals do after all, so why not let the robots follow their example? Climbing robots would indeed be an elegant solution. But such robots run the real risk of falling out of the trees, which is something that must be avoided as part of the survivability motif. One option could be to engineer the environment slightly so that the robots can stay close to the strategies employed by climbing, arboreal animals, yet remain safely suspended in the tree canopies for long periods of time. For these reasons, we choose to approach the question of how to perform environmental monitoring tasks in the tree canopies by letting the robots be wire-traversing, i.e., having them move along wires mounted between the trees. In this section, we discuss what such a design choice entails, followed by an experimental study based on the proposed design.[8]

8.2.1   Design Considerations

Wire-traversing robots are basically any types of robots capable of moving along cables, wires, and similar structures [303, 410], which makes them suitable for applications beyond environmental monitoring, such as agricultural robotics [51], or maintenance in hazardous locations, as in the case of power line inspection, e.g., [340]. In fact, this latter application has been a significant catalyst for the development of different styles of wire-traversing robots, e.g., [295, 336, 368]. Across the wire-traversing design spectrum, the common features include simplicity of the overall system design and, consequently, of the motion control, relatively small localization errors and navigation complexity—planning over a one-dimensional wire is not all that complicated—and low energy needs [294, 410].

These common features typically arise as direct consequences of the mechanical design, which plays such an important role in wire-traversing robots that we have to discuss this aspect of the SlothBot, as seen in Figure 8.2, in some detail. But this design cannot be considered in isolation, i.e., decoupled from the envisioned deployment scenarios. In fact, in order for the SlothBot to traverse a sufficiently wide area, a single wire is not going to be enough. For example, in precision agriculture (or in the tree canopies), a wire-traversing monitoring robot has to be able to traverse a mesh of wires and, therefore, to overcome the complications associated with wire-crossings.

---

[8]It should be stressed that this design choice is not the only possibility. Rather, it is one possible such choice that allows for safe, long-term operation among the trees.

Figure 8.2: Left: The SlothBot is a lightweight, solar-powered, minimally-actuated, wire-traversing robot, capable of switching between branching wires. Right: The SlothBot, dressed in its "sloth" shell, on long-term deployment at the Atlanta Botanical Gardens.

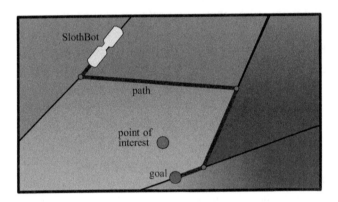

Figure 8.3: Example of an agricultural monitoring application using a wire-traversing robot. The different areas represent different crops in the field. The SlothBot is tasked with collecting measurements at a point of interest in the interior of the field. It, therefore, has to traverse the wires along a path and negotiate intersections, until it reaches the goal point, i.e., the point on the wire-mesh that is closest to the point of interest.

To stress the need for wire-switching capabilities, Figure 8.3 depicts a possible scenario playing out in an idealized agricultural setting. The different areas correspond to different crops in the field, while the lines represent a mesh of wires. The objective is to monitor relevant phenomena that evolve over long time-scales, such as crop growth. For this task to be successfully completed, *energy efficiency* and *fail-safeness* are required features, so that maintenance requirements and risks of failure are minimized. Moreover, the robot has to be able to negotiate wire intersections. For this to be possible, the robot has to have the capability to switch between different wire branches, without losing its grip.

The takeaway from this detour is that the SlothBot must be capable of moving on a mesh of wires, while performing long-term environmental monitoring tasks. To ensure that it is self-sustaining from an energy vantage point, it also needs to be equipped with solar panels for autonomous charging. The mechanical design should moreover be comprehensible and compact, while at the same time ensure that the robot remains safely attached to the wires at all times, even during the switching maneuvers.[9]

As fail-safeness is critical to survivability, and falling from the treetops to the forest floor below is not only mission-ending, but also an existential threat to the robot, we start the discussion here. For the purpose of wire-traversing, environmental monitoring robots, "fail-safe" must, first and foremost, mean that the robot does not fall down—it should remain on the wire at all times.

Note that the requirement of remaining safely suspended at all times has to hold true even if the actuators fail. Were this to happen, the SlothBot would essentially end up being transformed from a mobile monitoring robot to a stationary one. But, just because it cannot move, it does not mean that it should stop doing something useful; this is completely analogous to what happened to the Mars rover *Spirit* as it got stuck in a Martian sand pit, transforming from a mobile rover to a stationary science platform [421].

Table 8.1 shows a comparison between the eventual, but not yet fully revealed, SlothBot and other wire-traversing robotic platforms. The metrics used in the comparison are the wire-switching capabilities, the fail-safeness (as defined in terms of remaining on the wire despite actuator failures), and the actuation complexity, as measured by the number of actuators. Although not necessarily an entirely critical metric, the robot weight is also included, as lighter is typically better when it comes to a successful existence among the treetops. As seen, the SlothBot possesses all the desirable characteristics for

---

[9]Surveys of robots designed to traverse cables can be found in [294, 410]. Additional designs are provided in [18, 82, 286, 340, 348], including a robot that can "fold" itself around obstacles [340], a modular robot that slides on horizontal wires and climbs up vertical ones [348], caterpillar-like locomotion strategies for climbing up ropes [82], and a robot that is able to locomote using different modus operandi, such as inchworm-like or brachiating motion patterns [286].

	Locomotion	Wire-Switching	Fail-safe	No. Actuators	Weight (kg)
LineScout [340]	Wheels	No	Yes	—	100
Caterpillar Robot [82]	Wheels	No	Yes	—	—
SkySweeper [286]	Pulley Arms	No	Yes	3	0.466
Expliner [112]	Wheels	Yes	No	6	60
Modular Robot [18]	Wheels	Yes	Yes	16	10
SlothBot [303]	Wheels	Yes	Yes	7	1

Table 8.1: Comparison between the SlothBot and other wire-traversing robot designs. (The missing entries denote that the corresponding information is not publicly available.)

it to be able to participate in long-duration deployments, and has the smallest number of actuators among the platforms capable of wire-switching—the particulars of the SlothBot design is the subject of the next section. For now, to summarize, the drivers behind the SlothBot design are:

1. Energy efficiency;
2. Fail-safeness; and
3. Wire-switching capability.

8.2.2 Mechanical Design

To strike the appropriate maneuverability-complexity balance, the SlothBot is comprised of two bodies, connected by an actuated hinge, as seen in Figure 8.6. Each individual body, as depicted in Figure 8.4, houses a driving motor in the middle of the body, connected to a rim on which a tire is mounted. The use of wheels for locomotion is simple, energy-efficient, and makes the SlothBot significantly more safe, as compared to brachiating (swinging) solutions.

The switching maneuver is made possible through four pairs of spur gears—two per body, as seen in Figure 8.4. Each individual pair is stacked vertically, with the top gear having a circumferential gap of 20°, making it look like a C-shape, as shown in Figure 8.5. The bottom gear is driven by a servo-motor, which allows for the gap to be oriented to three different positions, as illustrated in Figure 8.5 as well. These positions are Top, Left, and Right, which correspond to the robot going straight, turning right, and turning left, respectively. This novel wire-switching mechanism minimizes the required actuation to only one servo-motor per gear pair, thus significantly increasing the simplicity and compactness of the design, as required.

Figure 8.4: One of the two bodies comprising the mechanical part of the SlothBot. Source: Fig. 3 in Notomista et al. (2019).

Figure 8.5: The switching mechanism for the SlothBot. The top components of the robot always remain above the wires, while the bottom components are confined to stay below them. The C-shaped gear allows one of the wires—but never both—to disengage from the robot during a wire-switching maneuver. Source: Fig. 7 in Notomista et al. (2019).

The fail-safeness of the SlothBot is guaranteed through the use of the two bodies connected by a hinge consisting of a rotational joint, whose axis lies in the longitudinal plane of the robot, and is actuated by its own, dedicated servo-motor. The servo-motor ensures the alignment of the bodies with respect to the wire branch that is being traversed. Additionally, the wire is compressed between the tires and the top lids of both bodies, as per Figure 8.4. This arrangement ensures that the friction force remains high enough to allow the tires to move the robot.

The wire-switching mechanism employed by the SlothBot is, as required, robust against actuator failures [303]. More specifically, the SlothBot is designed in such a way that it firmly remains on the wire in case the actuators

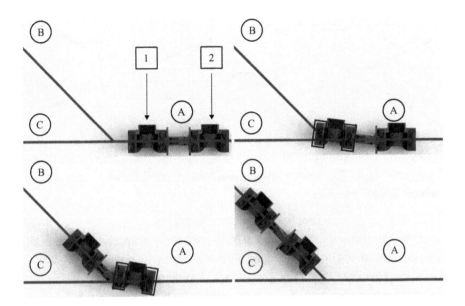

Figure 8.6: Wire-switching maneuver where the SlothBot switches from wire Branch A to Branch B. Source: Fig. 6 in Notomista et al. (2019).

fail during a wire-switching maneuver, or if the motors are actuated at the wrong time, i.e., a switch is attempted when there is no wire to switch to.

To further elucidate how this switching-mechanism works, Figure 8.6 shows a sequence of actions performed by the SlothBot in order to switch between different wire branches. In particular, the sequence involves the following steps:

- Figure 8.6 (Top Left): Both bodies of the SlothBot are on the same wire, indicated by Branch A, and the servo-motors keep all the gaps pointing straight up, thereby holding the top lids on the wire, while not allowing the wire to disengage. The objective is to switch from Branch A to Branch B;
- Figure 8.6 (Top Right): Body 1 is at the junction between the wires, and its gaps are both open, thereby allowing Branch C to disengage from Body 1;
- Figure 8.6 (Bottom Left): Body 2 undergoes the same maneuver as Body 1 just completed, thereby disengaging from Branch C and moving onto Branch B; and
- Figure 8.6 (Bottom Right): Both bodies are on the same wire, and the SlothBot has successfully switched from Branch A to Branch B.

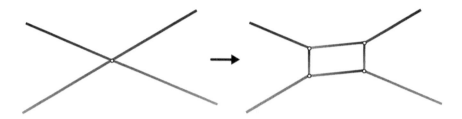

Figure 8.7: Example of turning one 4-way crossing into a sequence of four 3-way crossings. This modification is required as the SlothBot is only able to traverse 3-way crossings.

Due to the fail-safeness constraint on the switching maneuver, the SlothBot does not have the ability to negotiate wire-crossings with more than 3 branching wires. However, this is not a substantial limitation since any crossing can be turned into a sequence of 3-way crossings, as shown in Figure 8.7. Moreover, as can be deduced from Figure 8.6, the SlothBot cannot traverse crossings when the turning angle is smaller than 90°. However, this situation can always be avoided by performing a two-step maneuver in which two obtuse-angle crossings are negotiated instead of one acute-angle crossing [303].

With the mechanical design principles established, the next step is to realize this wire-traversing robot design in hardware. To this end, the SlothBot is constructed using rapid prototyping techniques, and all the main components are 3D-printed using standard PLA materials.[10] Two solar panels (visible in Figure 8.2), mounted on the sides of the SlothBot, are used to recharge the battery when the light intensity is high enough. A charging circuit is used to regulate the charging current based on the solar cell characteristics. This way,

---

[10]The printing time of an entire SlothBot is about 30 hours using a commercial 3D printer, while the assembly time is around 30 minutes, as all the other components are off-the-shelf. The resulting robot is 25.5cm long, 11.2cm high (13.5cm with the solar panels attached), and 6cm wide (31.2cm with the solar panels). The two wheel motors operate at 6V, with a 1000:1 reduction ratio, a maximum speed of 32rpm, and maximum torque of 0.88Nm, which allow the SlothBot to move at a leisurely maximum speed of 5cm/s. Additionally, the servo-motors used to rotate the spur gears are standard 9g servo-motors, with an operating voltage of 5V, and a maximum torque of 0.16Nm. The servo-motor used to actuate the hinge between the two bodies, required during the wire-switching maneuver, has a maximum torque of 1.52Nm, while operating at 5V. Taken together, the robot is powered by a rechargeable 7.4V, 1000mAh LiPo battery. Finally, an IoT-enabled microcontroller is used to control all 5 servo-motors and, through a dedicated motor controller, the 2 DC motors that drive the wheels.

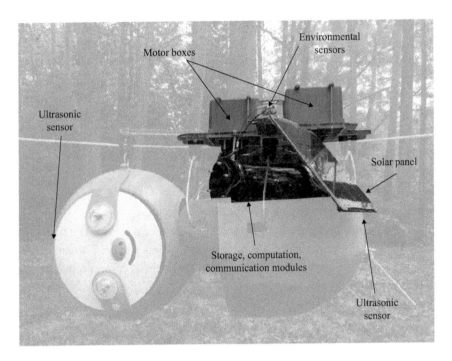

Figure 8.8: The hardware configuration of the SlothBot, where the outer, sloth-looking shell is faded in order to explicitly call out the hardware components that comprise the SlothBot.

the power efficiency, expressed as the ratio between the power that is transferred to the battery and the power received from the sun is maximized at each time instant.

As a final note, the SlothBot is designed to carry sensors for environmental monitoring, and the microcontroller connects to the sensors using the I^2C protocol, usually available on sensor data acquisition boards. Moreover, the microcontroller hosts a web server which handles requests for sensor data by a client running on a remote computer, responsible for storing the collected measurements, thus enabling remote environmental monitoring. These design choices are illustrated in Figure 8.8.

8.3   The SlothBot

Now that we have designed a robot that is capable of traversing wires and negotiating intersections, it needs to be put to use. To this end, imagine a wire-mesh mounted in the canopies in a forest ecosystem. To ensure longevity, i.e., to achieve persistent environmental monitoring, the survivability constraint, $Sh_{survive}[x; u] \geq 0$, in Equation 8.1 must hold. And, most of the time, the robot will not be doing anything in particular from a mobility point of view, i.e., the energy cost, $\|u\|^2$, is what is being minimized, as per Equation 7.28. But, in order to ensure that the robot can properly monitor an extended area, every now and then it has to move to a new location, as pictorially illustrated in Figure 8.3, which means that the minimum-energy controller must be replaced by a more purposeful, goal-oriented controller.

8.3.1 Motion Planning and Control

What is needed to ensure that the SlothBot can not only traverse wires in a safe but also purposeful manner is the development of a control law that will allow the robot to move on the wires to monitor events and collect data in the environment. This control law will serve as the nominal control input, u_{nom}, in the minimization problem,

$$\min_{u} \|u - u_{nom}\|^2, \tag{8.6}$$

that is to be solved, subject to the mandatory survivability constraint. In this section, we discuss how this nominal control input can be produced.

Assume, without loss of generality, that the mesh of wires is planar and that the layout is known to the robot. Let $X \subset \mathbb{R}^2$ be a closed and convex polygon, representing the environment in which the robot is deployed. Provided that there is no slip between the driving tires of the robots and the wires, we can assume that, by acting on the torques of the two DC motors, one can directly regulate the robot to a desired velocity. As such, a simplified robot model is given by the familiar, single integrator model,

$$\dot{x} = u, \tag{8.7}$$

where $x \in X$ is the robot position, and $u \in \mathbb{R}^2$ is the velocity control input. Note, however, that the robot cannot actually move all over X, as its motion is restricted to the wire-mesh, i.e., the model in Equation 8.7 must be constrained to reflect this fact.

The environmental monitoring task for the SlothBot is related to the problem of driving the robot as close as possible to a (possibly time-varying) goal location, $p(t) \in X$, in order to collect relevant measurements.

Motion-planning algorithms for wire-traversing robots typically use a graph or road-map search, e.g., [225], to find the path to a given location on the graph. Once the route through the graph is established, the robot can then be driven by a lower-level controller to the goal. However, in order to derive a nominal, continuous-time control law that will continuously drive the robot to the point on the wires closest to $p(t)$, this standard approach must be modified somewhat.

To arrive at the nominal controller, u_{nom}, consider the following constrained optimization problem,

$$\min_{q \in \mathscr{G}} \frac{1}{2} \|q - p(t)\|^2, \tag{8.8}$$

where $\mathscr{G}$ denotes the mesh of wires, or, more precisely, the set of points in X that belong to the line segments defined by the wires. In [301, 305], a motion control algorithm was provided that continuously maps the velocity of an unconstrained robot onto the velocity of a robot constrained to move on a wire-mesh. Armed with such a map, one can instead consider the negative gradient flow to the much simpler, *unconstrained* optimization problem

$$\min_{\rho \in X} \frac{1}{2} \|\rho - p(t)\|^2, \tag{8.9}$$

which is given by the proportional controller

$$\dot{\rho} = -\frac{1}{2} \frac{\partial}{\partial \rho} \|\rho - p(t)\|^2 = p(t) - \rho. \tag{8.10}$$

At this point, one can follow the process laid out in [301] directly and map $\dot{\rho}$ onto the wires to obtain the requisite $\dot{q}$ that ensures that q evolves in such a way as to remain in $\mathscr{G}$. The expression for the mapping from the unconstrained velocity, $\dot{\rho}$, to the constrained one, $\dot{q}$, as developed in [301], takes on a significantly less complex form when the robot is constrained to move on triangle meshes [303], which is the type of mesh utilized by the SlothBot, as per a previous discussion. The resulting, nominal control law is given by the expression

$$u_{nom} = \dot{q} = \widetilde{M}(\dot{\rho}), \tag{8.11}$$

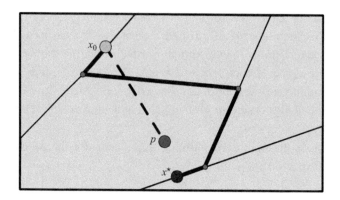

Figure 8.9: The robot, initially at x_0, is tasked with collecting measurements at point p. Due to the wire constraints, the robot instead moves to the location on the wire-mesh that is closest to p, denoted by x^*, by executing the controller in Equation 8.11.

where $\widetilde{M}(\dot{p})$ is the velocity, $\dot{p}$, mapped onto the wires, and the mapping $\widetilde{M}(\cdot)$ is the so-called *continuous-onto-wires* mapping from [301], which depends on the geometry of the environment and the arrangement of the wires.[11]

Putting everything together gives the actual control input, u, as the solution to

$$\min_u \|u - u_{nom}\|^2$$
$$\text{subject to } S_{h_{survive}}[x; u] \geq 0, \tag{8.12}$$

where u_{nom} is given in Equation 8.11, and where the survivability constraint is assumed to act exclusively in directions that keep the robot on the wires.[12]

Figure 8.9 depicts the monitoring scenario outlined in Figure 8.3, but formulated in terms of the relevant quantities from the current discussion. The robot, initially positioned at x_0, has to go and collect measurements at the location denoted by p. Since it is constrained to move on the wires, a continuous projection of the unconstrained path (dotted line) onto the wires is generated

---

[11]The derivation of $\widetilde{M}(\cdot)$ and its precise mathematical structure is given in [301], and we omit these here for the sake of the narrative flow.

[12]If the constraints are allowed to force the robot away from the mesh, an additional application of $\widetilde{M}(\cdot)$ is required, e.g., $u_{actual} = \widetilde{M}(u)$, where u is the solution to the constrained optimization problem in Equation 8.12.

using Equation 8.11, resulting in the constrained path (solid line). Following this projected path, the robot arrives at the location x^*, as prescribed.

In order to ensure that the full-blown survivability constraint in Equation 8.1 holds, a dynamic model is needed. The motion of the SlothBot is, as we have just seen, largely unproblematic from a modeling point of view, since at its low operating speeds, a single integrator model suffices. However, this is not enough when it comes to the energy dynamics. In particular, the constituent homeostasis constraint in Equation 8.2 is given by the energy constraint

$$h_{homeostasis}(x) = (E_{max} - E)(E - E_{min}) \geq 0, \tag{8.13}$$

where E is the energy level of the SlothBot's battery.

Following the developments in Chapter 5, we model the battery dynamics as

$$\dot{E} = F(x, E, t) = k\left(w(x, E, t) - E\right), \tag{8.14}$$

where $k > 0$ is a charge/discharge gain, and the energy drain is given by

$$w(x, E, t) = \frac{1}{1 + \frac{1-E}{E} e^{-\lambda(I(x,t) - I_c)}}. \tag{8.15}$$

Additionally, the solar intensity map is given by $I : \mathcal{E} \times \mathbb{R}_+ \rightarrow [0, 1]$, which provides a description of the availability of energy resources in the environment.

The energy dynamics coefficients, k and λ, as well as the intensity model, I, must be established empirically, which was done for the SlothBot through repeated 24h experiment cycles, as seen in Figure 8.10. The reason for the 24h span is that a single battery charge can operate the SlothBot for around 12h without recharging, and the experiments must capture the charging as well as the discharging dynamics. Included in Figure 8.10 is the solar intensity, as recorded by the SlothBot's on-board luminosity sensor, which is part of its environmental monitoring sensor pack.

8.3.2 Long-Duration Deployment

In May 2020, the SlothBot was deployed among the trees as a new feature at the Atlanta Botanical Garden's "Canopy Walk," as seen in Figure 8.11 (Left). Equipped with temperature, humidity, barometric pressure, solar radiation,

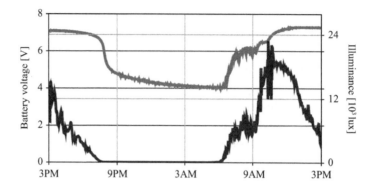

Figure 8.10: Battery voltage (the top line) and light intensity data (the bottom line) collected by the SlothBot at a particular location during a 24h cycle for the purpose of establishing the required environment-to-battery charge model in Equation 8.14. This, in turn, is needed for the homeostasis constraint in the overall survivability constraint in Equation 8.1.

Figure 8.11: Left: The SlothBot, suspended among the trees, as seen from the Canopy Walk. Right: Visitors to the Atlanta Botanical Garden are encouraged to look for the SlothBot.

luminosity, and carbon dioxide sensors, its task was to monitor the microclimate at this particular site. As the Atlanta Botanical Garden has an extensive plant conservation research operation, the idea was that the data generated would feed into the ecological niche models developed for the ecosystem in the botanical garden.

Figure 8.12: The SlothBot is deployed so as to be partially hidden from view, thereby evoking the feeling of the robot being a part of the ecosystem, rather than just being a high-tech "visitor" to the botanical garden.

Environmental monitoring was not, however, the SlothBot's only charge. As it was expected to perform its task over extended periods of time, long-duration autonomy was also on full display in a highly public location—the Atlanta Botanical Garden has over 700,000 visitors annually. As such, this particular deployment touches upon two of the central tenets of this book, namely long time-scales and tight connections between robot and environment.

The Canopy Walk is an elevated walkway, meandering its way through the treetops at the Atlanta Botanical Garden, and to stress the robot ecology theme, it was important that the SlothBot's operating range did not make it feel to the visitors like an alien piece of technology, artificially attached to the walkway, but rather as an almost organic part of the local ecosystem. To this end, the SlothBot was deployed 60ft up in the treetops, recessed 100ft away from the Canopy Walk. As a result, the SlothBot was partially obstructed away from view (see Figure 8.12), in a purposeful manner, hidden by the foliage at times, and visible at other times, depending on whether or not the homeostasis constraint would force the robot to move away and recharge. To ensure that

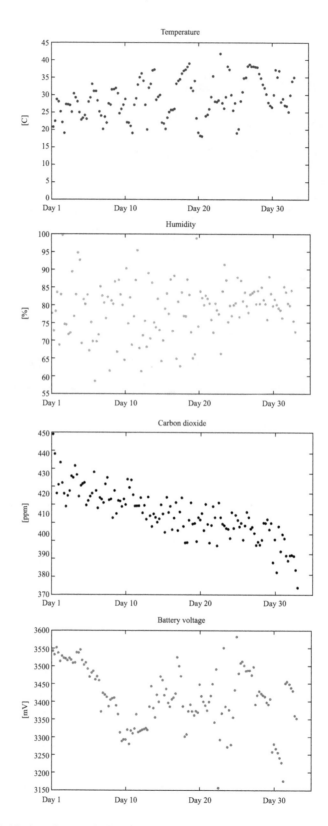

Figure 8.13: (caption on facing page)

the SlothBot's secondary, public outreach mission was not impeded by these arrangements, signage—shown in Figure 8.11 (Right)— encouraged visitors to be on the lookout for the elusive SlothBot.

The data collected during the first month of deployment is shown in Figure 8.13. As a stand-alone, this data is not necessarily all that instructive. But, when combined with the types of ecological niche models developed by conservation biologists, *the hope is that the data provided by the SlothBot, as well as by other environmental monitoring robots, will be able to contribute to the protection of Earth's diverse species and ecosystems, by having a persistent presence in the tree canopies.*

Figure 8.13: A subset of the microclimate data collected during the first month of the SlothBot's deployment in the Atlanta Botanical Garden. The data are temperature (Top), humidity (Second from Top), carbon dioxide levels (Second from Bottom), and battery levels (Bottom). Taken together, these types of data can help conservation biologists create the correlative models needed to make predictions about future environmental effects.

9 Autonomy-on-Demand

One of Brooks' design principles for building "Creatures" ("A Creature should do something in the world; it should have some purpose in being" [64]) stressed the fact that the robots should be *useful*. And, just because a robot is out on a long-duration deployment, it does not get a free pass in terms of usefulness. In fact, already in Chapter 2, we introduced a long-duration autonomy twist on the popular Laws of Robotics. While the First Law of Long-Duration Autonomy was "Don't die!" the Second Law stated that the robots should "Do something useful!"

Throughout this book, we have stressed survivability and minimized energy expenditure as playing central roles, i.e., we focused primarily on the First Law. One could even argue that we have embraced an extreme existential nihilism for the robots in that their existence is meaningless in terms of "life goals." And when there have been goals, they have been turned into constraints. In this final chapter, we remedy this omission and consider the Second Law, i.e., the question of usefulness, as it should be understood in the context of robot ecology.

It is certainly true that robots, deployed over long time-scales, may very well spend a significant part of their time doing nothing as was the case with the SlothBot in the previous chapter. But, every now and then, they will be recruited and tasked with participating in missions. We call this setup *autonomy-on-demand*. In this chapter, we go into detail about how an autonomy-on-demand robot system could be organized. Although the proposed setup is quite general, it will be instantiated through a particular autonomy-on-demand platform, namely the Robotarium, which is a remotely accessible, swarm robotics testbed where, most of the time, the robots are docked at charging stations, doing nothing beyond ensuring that their batteries are charged and that they are ready to go. But every now and then, remote users upload code, recruit (subsets of) the robots, and run experiments. As such, the Robotarium constitutes a canonical manifestation of the autonomy-on-demand idea. Before we can describe this platform, we first have to recall the formalism, developed in Chapter 7, that will serve as the engine for producing safe, long-lasting, and *taskable* multi-robot systems.

9.1 Recruitable Robots

Consider a team of N robots, with states $x_i \in \mathbb{R}^n$, $i \in [N]$. Assume that the individual robot dynamics are control-affine, which means that the ensemble-level, combined system inherits the control-affine form,

$$\dot{x} = f(x) + g(x)u, \tag{9.1}$$

where $x = [x_1^T, \ldots, x_N^T]^T \in \mathbb{R}^{Nn}$ and $u = [u_1^T, \ldots, u_N^T]^T \in \mathbb{R}^{Nm}$, with the individual control inputs taking on values in $\mathbb{R}^m$. Additionally, assume that $f(x)$ and $g(x)$ are locally Lipschitz continuous over the relevant domain, and that each robot's individual state can be mapped onto position and energy outputs,

$$\begin{aligned} p_i &= \eta_P(x_i) \in \mathbb{R}^d \\ E_i &= \eta_E(x_i) \in \mathbb{R}, \end{aligned} \tag{9.2}$$

where $d = 2$ for planar robots, and $d = 3$ for aerial robots.

9.1.1 Task Specifications

Under these reasonably mild assumptions, *survivability*, as the constraint that must always hold, is given by

$$h_{survive}(x) = \min\{h_{resources}(x), h_{interact}(x), h_{homeostasis}(x)\} \geq 0, \tag{9.3}$$

as per Equation 7.24. The corresponding, *actionable* CBF constraint, expressed as a linear constraint in the control signal, u, is given by

$$S_{h_{survive}}[x; u] \geq 0, \tag{9.4}$$

or, expanded out, as

$$L_f h_{survive}(x) + L_g h_{survive}(x)u + \alpha(h_{survive}(x)) \geq 0, \tag{9.5}$$

for some locally Lipschitz continuous, extended class $\mathcal{K}_\infty$-function, α. Note that once again the Boolean composition in Equation 9.3 ($h_1 \wedge h_2 \geq 0 \Leftrightarrow \min\{h_1, h_2\} \geq 0$) means that the Lie derivatives in Equation 9.5 are set-valued, as per Chapter 6.

In between missions, the robots are simply asked to expend as little energy as possible, i.e., to solve

$$\begin{aligned} \min_u \ &\|u\|^2 \\ \text{subject to} \ &S_{h_{survive}}[x; u] \geq 0. \end{aligned} \tag{9.6}$$

The autonomy-on-demand concept kicks in when the robots are asked to do something more than just survive. As discussed in the context of goal-driven behaviors in Chapter 7, two options are at our disposal as possible ways of encoding the task specifications. The first option is most closely aligned with the robot ecology principle of a "purposeful expenditure of energy," and it hinges on the requirement that the task performance can be captured by an instantaneous, positive definite cost, $J_{task}(x) \geq 0$.

As seen in Chapter 7 (Section 7.3), this is a requirement that a large number of multi-robot behaviors satisfy, including standard formation and coverage controllers, e.g., [93]. If the task is given in this form, the relevant barrier function, $h_{task}(x) = -J_{task}(x)$, can be constructed.

The resulting autonomy-on-demand formulation becomes:

$$\begin{aligned} \min_{u,\delta} \ &\|u\|^2 + \delta^2 \\ \text{subject to} \ &\begin{cases} S_{h_{survive}}[x; u] \geq 0 \\ S_{h_{task}}[x; u] \geq -\delta, \end{cases} \end{aligned} \tag{9.7}$$

where $\delta \in \mathbb{R}$ is the slack variable that captures how well the task is being executed, and

$$S_{h_{task}}[x; u] = L_f h_{task}(x) + L_g h_{task}(x)u + \hat{\alpha}(h_{task}(x)). \tag{9.8}$$

As before, $\hat{\alpha}$ is a locally Lipschitz continuous, extended class $\mathcal{K}_\infty$-function, and it is in general not equal to α in Equation 9.5. In fact, as discussed in Section 7.2.3, if

$$\hat{\alpha}(z) = cz^\gamma, \tag{9.9}$$

for some $c > 0$ and $\gamma \in (0, 1)$, the task is achieved in finite time, as opposed to asymptotically [306].

Although convenient and closely aligned with the robot ecology theme, such cost-based task formulations are not always possible to adhere to. In the Robotarium, users are free to upload whatever code they see fit, and there is simply no reason to believe that their intended tasks will conform to the prescribed form. Luckily, there is an alternative way of encapsulating user intent within the autonomy-on-demand framework. No matter how the task is specified, sooner or later it has to result in robot motions, i.e., in control signals for the robots to execute. As in previous chapters, we can let u_{nom} denote this nominal, user-specified control input. The idea is for the actual control signal to stay as close as possible to the nominal controller, without sacrificing survivability.

Formally, what this *minimally invasive* approach to long-duration autonomy entails is

$$\begin{aligned} &\min_u \ \|u - u_{nom}\|^2 \\ &\text{subject to} \ \ S_{h_{survive}}[x; u] \geq 0. \end{aligned} \tag{9.10}$$

This is the interpretation of autonomy-on-demand that is employed in the Robotarium. In the next section, we discuss the particulars of how this is structured, followed by a characterization of the Robotarium hardware/software design needed to support the control architecture. Additionally, we report on some of the autonomy-on-demand experiments that have been executed on the Robotarium.

9.1.2 Remote Access Control in the Robotarium

Throughout this book, statements have been made about general, control-affine systems on the one hand, and single integrator systems on the other. But real robots typically fall somewhere in between. As such, one must first figure out how to map the rather abstract formulation in Equation 9.10 onto real, physical robots.

The robots in the Robotarium are wheeled, differential-drive mobile robots [338], i.e., they are equipped with two independently controlled wheels of radius R, where the control inputs are the angular velocities of the right and left wheels, respectively. At the speeds at which the Robotarium robots operate, kinematics typically suffice, i.e., dynamical effects such as those associated with friction or inertial factors can largely be neglected.

Let the pose of Robot i be given by its position in the plane, (x_i, y_i), and its orientation, ϕ_i. The kinematics of the differential-drive robot is

$$\dot{x}_i = \frac{R}{2}(\omega_{r,i} + \omega_{l,i}) \cos \phi_i$$
$$\dot{y}_i = \frac{R}{2}(\omega_{r,i} + \omega_{l,i}) \sin \phi_i \qquad (9.11)$$
$$\dot{\phi}_i = \frac{R}{L}(\omega_{r,i} - \omega_{l,i}),$$

where $\omega_{r,i}$ and $\omega_{l,i}$ are the angular velocities of the right and left wheel, respectively, and L is the length of the wheel axis.

Although this is how the robots are actually controlled in the Robotarium, it is not, however, particularly natural or convenient to define motions in terms of something as nebulous as wheel velocities. To remedy this, the standard trick is to map the dynamics in Equation 9.11 to a unicycle model, where the control inputs are instead given by the translational and angular velocities of the robot, denoted by v_i and ω_i, respectively. The unicycle dynamics for Robot i, $i \in [N]$, is given by

$$\dot{x}_i = v_i \cos \phi_i$$
$$\dot{y}_i = v_i \sin \phi_i \qquad (9.12)$$
$$\dot{\phi}_i = \omega_i.$$

What is required in order to align these two models is a mapping from (v_i, ω_i) to $(\omega_{r,i}, \omega_{l,i})$ in order to be able to design for (v_i, ω_i), yet execute $(\omega_{r,i}, \omega_{l,i})$ on the actual robots.

By equating $(\dot{x}_i, \dot{y}_i, \dot{\phi}_i)$ in Equations 9.11 and 9.12, we get

$$v_i = \frac{R}{2}(\omega_{r,i} + \omega_{l,i})$$
$$\omega_i = \frac{R}{L}(\omega_{r,i} - \omega_{l,i}), \qquad (9.13)$$

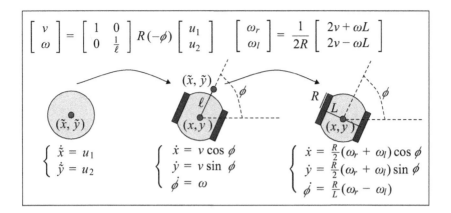

Figure 9.1: Progression from single integrator dynamics—for the purpose of formulating survivability constraints and user specifications—via an intermediary unicycle model, to the differential-drive model of the actual Robotarium robots.

which inverts to

$$\omega_{r,i} = \frac{2v_i + \omega_i L}{2R}$$
$$\omega_{l,i} = \frac{2v_i - \omega_i L}{2R}. \tag{9.14}$$

As a result, we can design controllers for the unicycle model, and then generate $(\omega_{r,i}, \omega_{l,i})$ from the velocities (v_i, ω_i).

One can certainly operate directly in (v_i, ω_i)-space. But, to tap in more directly to the many examples we have discussed throughout the book, where it is assumed that the robot positions can be directly controlled, one additional step is needed. In fact, if we are willing to ignore the orientation of the robots, this can be achieved by considering points off the wheel axis of the robots, offset by a distance $\ell > 0$, i.e., to consider the new points

$$\tilde{x}_i = x_i + \ell \cos \phi_i$$
$$\tilde{y}_i = y_i + \ell \sin \phi_i, \tag{9.15}$$

as shown in Figure 9.1.

Equation 9.12 dictates that the time derivatives of these new variables are given by

$$\dot{\tilde{x}}_i = v_i \cos \phi_i - \ell \omega_i \sin \phi_i$$
$$\dot{\tilde{y}}_i = v_i \sin \phi_i + \ell \omega_i \cos \phi_i. \tag{9.16}$$

If we now postulate that we can control this offset position directly, i.e., that $(\dot{\tilde{x}}_i, \dot{\tilde{y}}_i)^T = u_i \in \mathbb{R}^2$, we can invert the expression in Equation 9.16, as per [315], to get (v_i, ω_i) in terms of u_i as

$$\begin{bmatrix} v_i \\ \omega_i \end{bmatrix} = \begin{bmatrix} 1 & 0 \\ 0 & \frac{1}{\ell} \end{bmatrix} R(-\phi_i) u_i, \tag{9.17}$$

where $R(\cdot)$ is the rotation matrix,

$$R(\psi) = \begin{bmatrix} \cos\psi & -\sin\psi \\ \sin\psi & \cos\psi \end{bmatrix}. \tag{9.18}$$

If we, as a final construction, set $p_i = [\tilde{x}_i, \tilde{y}_i]^T \in \mathbb{R}^2$, we have the Robotarium robot dynamics on the familiar form, $\dot{p}_i = u_i$, with the additional energy dynamics, $\dot{E}_i$, given by Equation 7.1. The full transformation chain, from single integrator dynamics to differential-drive, wheeled robots, is given in Figure 9.1.

These mappings between models allow us to express the survivability constraints in terms of the requisite positions and energy levels. In fact, returning to the constraint in Equation 9.3, the particular variants of the constituent constraints that are enforced in the Robotarium are

$$\begin{aligned} h_{resources}(x) &= \min\{h_{obstacle}(x), h_{anchor}(x)\} \\ h_{obstacle}^{ind}(x_i) &= \min_{w \in \mathcal{W}} \{\|p_i - w\|^2 - D_o^2\} \\ h_{anchor}^{ind}(x_i) &= (E_i - E_{min}) - \mathcal{E}(dist(p_i)) \geq 0 \\ h_{homeostasis}(x) &= h_{energy}(x) \\ h_{energy}^{ind}(x_i) &= (E_{max} - E_i)(E_i - E_{min}) \\ h_{interact}(x) &= h_{separation}(x) \\ h_{separation}^{pair}(x_i, x_j) &= \|p_i - p_j\|^2 - D_s^2. \end{aligned} \tag{9.19}$$

The only novelty in these expressions, as compared to what was discussed in Chapter 7, is the inclusion of a term encoding avoidance of the Robotarium walls. To that end, $\mathcal{W}$ is the set describing the wall locations, where the charging stations are not considered part of the wall, i.e., $h_{obstacle}^{ind}$ ensures that every robot stays at least D_o away from the wall. At the same time, $\mathcal{E}(dist(p_i))$ is the energy required to travel to the closest, available charging station, and D_o

```
% Set the number of robots used in the experiment
N = 20;

% Set up the Robotarium object
r = Robotarium('NumberOfRobots', N, 'ShowFigure', 'true');

% Set the number of iterations for the experiment. Each time-step
% when deployed on the Robotarium is ~0.033s
iterations = 1000;    % ~30 second experiment

for i = 1:iterations

    % Get the current poses of all robots
    x_uni = r.get_poses();

    % Convert to single integrator dynamics
    x = uni_to_si(x_uni);

    %%% MAIN CODE GOES HERE %%%
    %%%
    %%% should produce velocities dx
    %%%

    % Map to unicycle control inputs
    dx_uni = si_to_uni_dyn(dx, x);

    % Set the robot velocities
    r.set_velocities(1:N, dx_uni);

    % Send the velocity commands to the robots
    r.step();

end

r.debug();    % Prints errors that can cause the submission to be rejected
```

Figure 9.2: Example code from the Robotarium API that contains an open spot where the actual control algorithm should be inserted—see Figure 9.3.

is selected rather carefully in such a way that the robots can always return and recharge. Armed with this survivability constraint, the robots are guaranteed to never collide with other robots, or with the walls. Additionally, they never get stranded away from a charging station with depleted batteries, as required.

That takes care of the constraints. What is missing is the least-squares cost in Equation 9.10. In the Robotarium, the user-specified, nominal control input for Robot i, $u_{nom,i}$, is generated by the uploaded control code, as shown in Figures 9.2 and 9.3. In Figure 9.2, the general code structure is presented, with an empty spot provided, where the main control code should be inserted. In Figure 9.3, two examples of such control codes are given for rendezvous (Top) and formation control (Bottom) applications.

```
for i = 1:N

    % Initialize velocity to zero for each agent.
    dx(:, i) = [0 ; 0];

    % Get the topological neighbors of agent i based on the graph
    % Laplacian L (must be previously specified)
    neighbors = r.getTopNeighbors(i, L);

    % Iterate through robot i's neighbors
    for j = neighbors

        %%% CONSENSUS %%%

        % For each neighbor, calculate appropriate consensus term and
        % add it to the total velocity
        dx(:, i) = dx(:, i) + (x(1:2, j) - x(1:2, i));

        %%% END CONSENSUS %%%

    end
end
```

```
for i = 1:N

    % Initialize velocity to zero for each agent.
    dx(:, i) = [0 ; 0];

    % Get the topological neighbors of agent i based on the graph
    % Laplacian L (must be previously specified)
    for j = r.getTopNeighbors(i, L)

        % For each neighbor, calculate appropriate formation control term and
        % add it to the total velocity

        %%% FORMATION CONTROL %%%
        % (requires weight functions)

        dx(:, i) = dx(:, i) + ...
        formationControlGain*(norm(x(1:2, i) - x(1:2, j))^2 - weights(i, j)^2) ...
        * (x(1:2, j) - x(1:2, i));

        %%% END FORMATION CONTROL %%%

    end
end
```

Figure 9.3: Rendezvous (Top) and formation control (Bottom) example codes to be inserted in the location indicated in the main code in Figure 9.2.

With that, the models and specification formalisms needed to map the Robotarium operations directly onto the autonomy-on-demand framework are in place. In subsequent sections, we describe the particulars of the Robotarium in greater technical detail, as well as provide examples of remote user experiments as a way to explicitly call out the flexibility afforded by a systematic and purposeful autonomy-on-demand process.

9.2 The Robotarium: An Autonomy-on-Demand Multi-Robot Platform

When the Robotarium officially opened up in August 2017, it issued an invitation to anyone in the world interested in multi-robot systems to upload code, free of charge. As such, the motivation was not to build an autonomy-on-demand showcase per se. Rather, it was designed to democratize access to a high-fidelity, multi-robot testbed for researchers, students, and educators around the world.

9.2.1 The Impetus Behind Remote-Access Robotics

In parallel with the developments of distributed control and decision algorithms for multi-robot systems, covered throughout this book and elsewhere, e.g., [68, 276], significant advances have been made also on the hardware side, including the miniaturization of the robot platforms themselves, more robust sensing and communication modalities, and an overall reduction in cost, e.g., [76, 199, 338]. Despite these advances, it is still cost-prohibitive for many researchers and educators to construct large-scale testbeds that reliably and repeatedly can manage tens to hundreds of robots simultaneously. Beyond the cost, maintaining a multi-robot testbed is complex as well as time-consuming. As a consequence, most research on distributed control of multi-robot systems is validated in simulation instead of hardware, as noted in [153, 214, 278, 404]. Yet, actual deployment is crucial to multi-robot research, as it is virtually impossible to faithfully predict and simulate all the issues associated with making multiple robots perform coordinated tasks in unstructured and even unknown environments. This is particularly true for robots deployed over long time-scales, as unexpected events will inevitably occur that were not part of the planned (and, subsequently, simulated) scenarios.

It is in this light that the Robotarium, shown in Figure 9.4, should be understood. It is an open, publicly available, remote-access multi-robot testbed, explicitly designed to address the *theory-practice gap*, by providing access to

Figure 9.4: Two versions of the Robotarium platform. Left: A coverage control algorithm is executed on an early prototype version of the Robotarium using 13 "GRITSBot" robots [338]. The desired density function is projected onto the testbed arena in the shape of the letter R. Right: The Robotarium testbed with the newer "GRITSBot X" robots [431], charging at the inductive charging stations surrounding the Robotarium. Note the nonsymmetrical markers mounted on top of the GRITSBot X robots used by the motion captioning system to uniquely identify the robots. Source: Fig. 1 in Pickem et al. (2017).

a testbed that is flexible enough to allow for a number of different scientific questions to be asked, and different coordination algorithms to be tested, thanks to the minimally invasive control strategy employed, as per Equation 9.10. An additional benefit of the type of access provided by the Robotarium is that it points towards the possibility of standardizing multi-robot testbeds, where algorithms can be tested, compared, and validated by users worldwide on the same computing, communication, and robot configurations.[1]

The Robotarium was explicitly designed to support *research*. And, as research is open-ended, sometimes messy, and oftentimes unsuccessful, this means that the Robotarium structure cannot overly restrict the types of experiments that can be submitted. But, at the same time, it must be safe, in the sense that robots cannot be allowed to crash into each other, or into the surrounding arena walls, as a result of an unsuccessful experiment, which is why the autonomy-on-demand formalism in Equation 9.10 is such a natural choice

---

[1]It should be noted that a number of other, remote access testbeds have been successfully deployed in their respective domains, such as other multi-robot testbeds [133, 198, 199, 282, 329, 418], sensor network testbeds [3, 119, 280, 352], and remotely accessible educational tools [76, 168, 335]. Additional pointers can be found in the surveys [168, 198, 409] and the references therein.

for encoding safety and task specifications in the Robotarium.[2] In the remainder of this section, we describe how the resolution to the flexibility/safety tension is structured and realized in more detail.

9.2.2 Testbed Design

Although the primary focus of the book is on control design and on the architectural issues surrounding the design choices required of long-duration deployments, any description of the organization of an autonomy-on-demand platform must involve at least some coverage of hardware and software issues. In particular, the continuous operation of the Robotarium relies on the ability to support automated experimentation, which in turn hinges on robust position tracking, automated battery recharging, and provably collision-free motions. We here outline how these requirements informed the design of the Robotarium.

The following design considerations must be taken into account, in accordance with [126, 337]:

- Inexpensive yet robust robots;
- Intuitive interaction modalities and easy-to-use data collection through a public web interface that supports code submission and data/video retrieval;
- Seamless transitions between development in simulation and execution on the physical robots, facilitated by a data-driven characterization of the simulation-hardware gap of the robots;
- Minimized cost and complexity of maintaining a large collective of robots through built-in features, such as automatic charging and tracking; and
- Integrated safety and security measures to protect the Robotarium from damage and misuse.

In subsequent paragraphs, we show how these requirements are addressed, organized around the themes of automated charging, position and display systems, and robot platforms.[3]

---

[2]A report from a National Science Foundation Workshop on Remotely Accessible Testbeds [126] identified the inherent safety/flexibility tension as one of the key issues when formulating a "science of remote access."

[3]In the discussion, we keep the hardware specifics general, as the particular choices are not material to the overall design philosophy. For the sake of completeness, however,

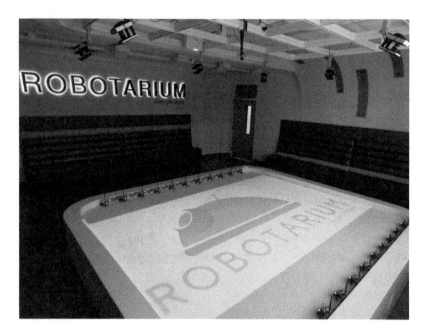

Figure 9.5: The Robotarium, situated inside a refurbished classroom at the Georgia Institute of Technology, featuring an elevated arena, a motion-capture system attached above the testbed, as well as wireless chargers installed on the arena walls.

Automated Charging

The Robotarium testbed, where the experiments are executed, is a 3.65m × 4.25m custom-made, elevated arena—see Figure 9.5. The walls of the testbed are outfitted with inductive chargers, shown in Figure 9.6, that allow the robots to autonomously recharge in between missions and during experiments where not all robots are required, or when low battery levels trigger a return to the charging stations. This setup keeps the robots available for use over long time periods, without the need for any human intervention. Wireless inductive chargers were chosen over conductive rail charging for general safety reasons, as conductive chargers always run the risk of being shorted, which

---

we here provide the specifics: The inductive chargers are Qi chargers; motion captioning is done using 8 Vicon cameras; the video camera is a 2 Megapixel ELP camera; and the overhead projector is an Optoma EH200ST projector.

Figure 9.6: The Robotarium's wireless charging system, consisting of trans-mitters on the wall of the testbed and receivers on the back of the GRITSBot X robots.

can be catastrophic for a testbed that is left unattended, while in operation, for long periods of time.[4]

Positioning and Visualization

A collection of motion capture cameras are mounted above the perimeter of the testbed in order to track each robot's motion for data acquisition and control purposes. Each robot is identified through a unique, nonsymmetrical pattern of markers, seen on the top of the robots in Figure 9.4 (Right). The motion capture system supports sub-millimeter precision pose information at a rate of 120Hz for each robot in the arena. From a hardware safety standpoint, the speed, accuracy, and precision of the motion capture system allows the Robotarium to detect potentially harmful situations during the experiments,

---

[4]Perhaps even more importantly, the Robotarium is a highly toured facility, where conductive charging has the potential to harm visitors who may accidentally bridge the charging leads.

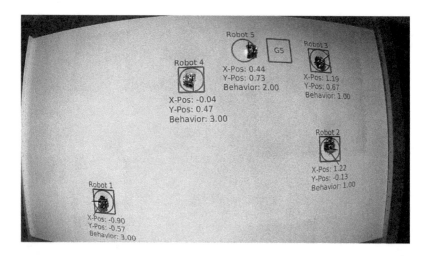

Figure 9.7: An overhead image of an experiment in the Robotarium. Projected are discrete behavior modes, robot identifiers, robot positions, and the robots' goal locations. Source: Fig. 3 in Wilson et al. (2020).

including the real-time evaluation of the survivability constraint in Equation 9.4, which can then be autonomously corrected and averted.

A video camera, used for automatic capture of the experiments, is mounted over the center of the testbed. This allows users to review their experiments in order to identify potential flaws or undesirable behaviors, as well as use the video to supplement and support their own scholarly research products. To convey additional information during the execution of the experiments, an overhead projector lets the users project time-varying, environmental backgrounds onto the testbed. Examples of user-generated scenes include urban cityscapes, forests with spreading wildfires, insides of buildings, ant colonies, farm fields, road networks, and even planetary vistas, just to name a few [431]. What can additionally be displayed are visualizations of control functions, such as potential fields or control vectors, or projections of other potentially useful information, such as battery voltages, network connectivity, or robot identifiers. Figure 9.7 shows an example usage of the projector, where robot goal locations are projected onto the testbed, along with other, relevant robot state information.

Robotic Platforms

During its first two years of operation, the Robotarium was populated with miniature, custom-made, differential-drive mobile robots, known as

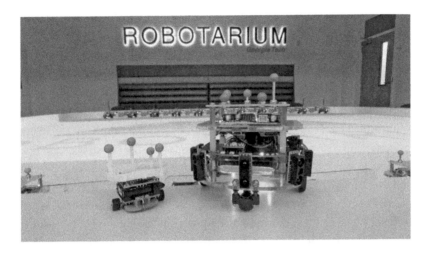

Figure 9.8: The Robotarium's original inhabitant, the GRITSBot (Left) [338], next to its updated, more robust replacement, the GRITSBot X (Right) [431].

GRITSBots [338]. These robots, pictured in Figure 9.8, feature a modular design, with autonomous charging and wireless reprogramming capabilities. However, like many miniature differential-drive robots, they face reliability issues when operating for long periods of time. For instance, the small, low-cost motors (especially at lower battery voltages) lack the necessary torques needed to overcome mechanical disturbances, e.g., from extraneous particulate or manufacturing inconsistencies in the testbed surface. This issue, in combination with a relatively limited battery life, prompted the development of a new robotic platform. As such, the successors to the GRITSBots are the larger and more robust, differential-drive platforms, the GRITSBot X robots (seen in Figure 9.8), which support more reliable, long-term operation [131, 431].

9.2.3 Safety and Robust Barrier Functions

As discussed in Section 9.1.2, safety and survivability in the Robotarium is ensured through the constituent constraints in Equation 9.19. Beyond adhering to the broader autonomy-on-demand theme, these constraints were selected with the following three design considerations in mind:

- All robots are provably safe in the sense that collisions are avoided;
- Users' commands are only modified when collisions are imminent; and
- Collision-avoidance is executed in real-time (in excess of a 30Hz update rate).

The first of the three bullets is (in theory) ensured directly through the forward invariance property associated with $h_{survive}(x) \geq 0$ in Equation 9.3, while the second bullet is covered by the minimally invasive modification of the user commands in Equation 9.10. The third bullet, however, requires some further scrutiny.

In Chapter 4, it was observed that the formulation in Equation 9.10 (originally introduced in Equation 4.20) is a quadratic programming problem in that the cost is quadratic in the decision variable, u, while the constraints are linear in u. From a computational complexity vantage point, this is mostly harmless and unproblematic [58, 252]. But there are some issues here that must be untangled when it comes to real-time execution. Although the number of decision variables, as well as the majority of the constraints in Equation 9.19, grow linearly in the number of robots in the team, the collision-avoidance constraint, $h_{separation}(x) \geq 0$, grows quadratically due to the pairwise nature of the constraint. A possible way to cap this growth is to only consider pairs of robots that are close enough for collisions to be potentially imminent, as was done in Chapter 4, following the developments in [56]. Coupling this approach with the observation that the robots' physical dimensions limit the maximum possible robot density, yields an upper bound on the number of pairwise interactions that need be considered in the collision-avoidance constraint.

In the Robotarium, the default, inter-robot safety distance is set to $D_s = $ 8cm, and the relevant neighborhood radius of 20cm is the radius of the area in which potential collisions must be considered. As a result, a generous upper bound on the size of any robot's neighborhood is given by at most 26 robots, which limits the size of each individual robot's separation constraint to no more than 26 pairwise constraints, regardless of the size of the overall team.

In Chapter 4 (Sections 4.3.1 and 4.3.2), centralized and decentralized versions of the collision-avoidance routine were introduced, and the decentralized version is what is ultimately deployed in the Robotarium. The computation times associated with both of these two varieties are shown in Table 9.1, where, for the sake of producing fair comparisons, the decentralized barrier functions are computed on the same central unit that computes the centralized CBFs.[5] Thus, the total computation time, T, is divided by the number of robots, N, in the decentralized case, to capture the computation time associated with the corresponding, parallel implementation. As Table 9.1 shows, while the centralized performance deteriorates as expected when scaling up

---

[5]The computations were done on an Intel I7 4790, 3.6GHz, with 16GB of memory.

Team Size, N	Centralized CBFs, T [ms]	Decentralized CBFs, T/N [ms]
10	5.6	3.2
40	11.6	3.5
100	78.0	5.4

Table 9.1: Computation times associated with solving the quadratic programming problem in Equation 9.10, subject to pairwise collision-avoidance CBF constraints, displayed as a function of team size.

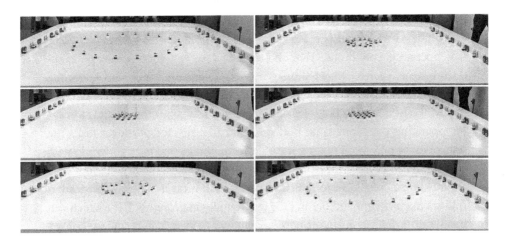

Figure 9.9: 20 GRITSBots are tasked with swapping positions by going through the same point in the middle of the Robotarium. The positivity of the barrier function, $h_{separation}$, is maintained throughout the maneuver. As a result, the robots are able to successfully complete the task without collisions.

the number of robots, the computation time in the decentralized case remains well below 10ms, even for 100 robots. In fact, the Robotarium can handle collision-avoidance for 100 robots, with an update frequency of 185Hz, and therefore scale gracefully in the number of robots without compromising the update rates [337].

That takes care of the third bullet in the list of safety requirements. An example of the decentralized barrier functions in action is shown in Figure 9.9, where 20 robots are asked to swap positions with each other by having all of them go through exactly the same point in the middle of the arena. In other words, the user-specified control code would, if left to its own devices, result in a massive, 20-robot "car crash." But, as can be seen, the safety barrier function kicks in as they get close to the intersection point, and they are able to safely navigate around each other and eventually achieve the requested position swap.

It is now tempting to declare success and rely entirely on the CBF formalism developed thus far. Unfortunately, one issue surrounding CBFs that they share with nearly all nonlinear control techniques is that their effectiveness tends to depend on the accuracy of the model. The statement that "$S_h[x; u] \geq 0$ implies that $h(x) \geq 0$, $\forall t \geq 0$," only holds true for the model, $\dot{x} = f(x) + g(x)u$, and may not necessarily hold for the actual system, depending on how closely it aligns with the model. In the Robotarium, the differential-drive model is reasonably accurate at moderate speeds and nominal operating conditions. But, this model does not incorporate potential disturbances, such as wheel slips or network latencies, which may cause the robots to collide, despite the theoretical guarantees provided by the CBF.

As the Robotarium is a long-duration autonomy platform, i.e., it is always on, operating over long time horizons, and is ready for whatever tasks the remote users may want the robots to perform, things that *may* go wrong *will*, indeed, sooner or later go wrong. To overcome this issue, the Robotarium goes slightly beyond the CBF formalism developed thus far in the book. Instead, it follows the program laid out in [131], by employing *robust* CBFs, which can handle disturbances in the robots' dynamics, modeled as bounded, convex sets. By relying, once again, on the theory of differential inclusions, one may guarantee that all possible, disturbed solutions remain collision-free.

In order to capture physical disturbances and deteriorating network effects, but without having to try to model these effects in great detail, the basic method employed in the Robotarium to achieve robustness when it comes to satisfying the survivability constraints, is to augment the system dynamics with a disturbance, i.e.,

$$\dot{x} \in f(x) + g(x)u + \text{co}\{\Psi(x)\}, \qquad (9.20)$$

where $\Psi(x) \subset \mathbb{R}^n$ is the disturbance set that captures the set of possible (or observed) disturbances at x, and $\text{co}\{\cdot\}$ denotes the convex hull. As shown in [131], this set can be learned over time, as the robots move around in the Robotarium.

Just as was done in Chapter 6 to manage set-valued Lie derivatives, the fact that the dynamics in Equation 9.20 are given by a differential inclusion,

Figure 9.10: A group of 7 GRITSBot X robots completes an iteration of a repeated robot-robot position swap experiment. The robots are initially arranged
on a circle (Left) and attempt to traverse to the opposite side, while utilizing
the robust version of the CBF formalism in Equation 9.22 in order to prevent
collisions (Middle), and ensure that each robot safely reaches the opposite
side of the circle (Right).

rather than by a differential equation, means that the CBF constraint needs
to be modified to ensure that the proper inequality holds across *all* possible
values rather than at a singleton. In other words, the constraint goes from the
familiar form

$$\frac{\partial h^T}{\partial x}(f(x) + g(x)u) \geq -\alpha(h(x)), \tag{9.21}$$

to an inequality over the worst-case scenario,

$$\min\left\{\frac{\partial h^T}{\partial x}\left(f(x) + g(x)u + \text{co}\{\Psi(x)\}\right)\right\} \geq -\alpha(h(x)), \tag{9.22}$$

where the minimum is taken over the convex hull of the disturbance set, $\Psi(x)$.
     With this slight modification, survivability in the Robotarium can be en-
sured in a robust manner. To appreciate why the robustification modification is
needed, consider again the scenario depicted in Figure 9.9, where the robots
are asked to swap positions by going through the same point in the middle
of the arena, thus being on a guaranteed collision-course in the absence of
the survivability constraint. Rather than performing this swap just once, and
in order to explicitly call out the efficacy of the robust CBFs, this robot-robot
position swap is repeated a number of times, as shown in Figure 9.10, with and
without the disturbance set added to the dynamics. In Figure 9.11, $h_{separation}$
is plotted for both of these cases. As can be seen, the robust version ensures
safety for all times, while the non-robust version suffers from non-safe situations, where $h_{separation}$ becomes negative during the experiment, as per [131].

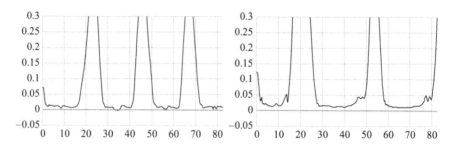

Figure 9.11: Plots of the smallest of the constituent, pairwise separation constraints used to compose the collision-avoidance constraint, $h_{separation}$. Depicted are the values without (Left) and with (Right) the robust CBF modification, respectively. If, at any point in time, the value falls below zero, the collision-avoidance constraint is violated. As can be seen, the robust CBF formulation (Right) experiences no such situations; whereas the non-robust CBF formulation (Left) does indeed encounter such a constraint violation.

9.3   Remote Experimentation

An autonomy-on-demand platform is ultimately only successful if the "demand" part is realized, i.e., if people are actually using it. And usage depends on a number of factors, such as exposure, ease-of-use, and the flexibility and breadth of the types of experiments that can be supported by the platform.

The way remote users interact with the Robotarium is nominally to start out with the Robotarium simulator. Once the experiment runs successfully on the simulator, the code is submitted through the Robotarium website.[6] Following a *first-in, first-out* queueing rule, the Robotarium server retrieves the next submission in the queue and runs a rudimentary safety-verification procedure on the code [337]. If the experiment passes this check, the Robotarium deploys the required number of robots, executes the experiment code, and collects the requested data, such as pose information or video of the experiment. Afterwards, the users are notified as to the experiment status, and are given experiment data access. More detailed explanations of this workflow and the submission process are provided in subsequent paragraphs.

The Robotarium was made publicly accessible, free of charge, for educational and research purposes, in August 2017. Since then, thousands of

---

[6]The Robotarium simulator runs either Python or MATLAB, and the Robotarium website is www.robotarium.gatech.edu.

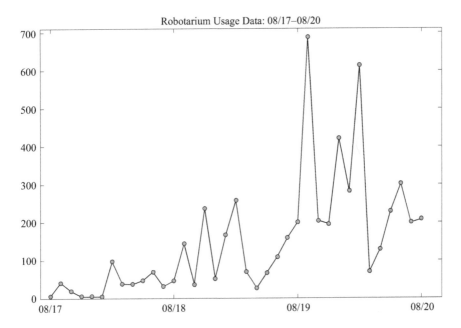

Figure 9.12: The number of experiments executed on the Robotarium each month during a three-year period since its official launch in August 2017.

experiments have been supported, and the number of experiments executed each month are shown in Figure 9.12 for the three-year period following its launch. As can be seen, the overall usage has grown significantly since the initial, somewhat slow start, to many hundreds of submissions each month. Over this three-year period, over 1500 distinct users from every continent (except Antarctica) have created Robotarium accounts, and have collectively executed over 5000 experiments, resulting in almost 100 peer reviewed publications that use the Robotarium as the generator for the experimental results. As such, it is fair to say that the Robotarium has delivered on its promise to democratize access to a high-fidelity, multi-robot platform.

Figure 9.12 calls out the fact that the number of experiments executed exhibits high month-to-moth variability. Further analysis, reported in [431], reveals that the overall submission trend is indeed growing, and that the spikes correspond to major robotics conference submission deadlines or end-of-semester class projects.[7]

---

[7]Note that the large drop in the number of experiments in March 2020 was likely due, in part, to the Covid-19 pandemic as many schools suddenly closed their campuses, and schedules were disrupted.

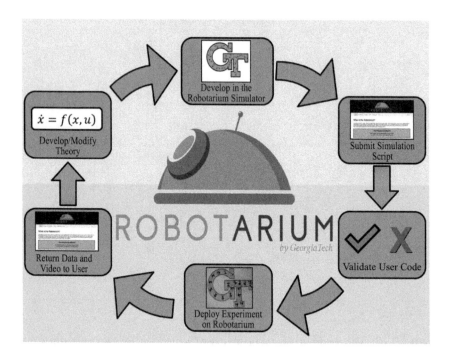

Figure 9.13: The Robotarium submission process. Source: Fig. 5 in Wilson et al. (2020).

9.3.1 Submission Process

Users access the Robotarium through a web interface, which includes links to the Robotarium simulation API repositories. In order for code to be executed by the robots on the physical Robotarium testbed, users must first create an account,[8] which allows them to submit files through the website, track the status of their submitted experiments, receive e-mail updates when their experiments are completed, view the data and video returned from their experiments, and interface with the Robotarium support team.

After creating an account, users enter the development cycle depicted in Figure 9.13. Their control algorithms must first be implemented in the Robotarium simulator, and when they are satisfied with the performance of their simulated experiments, their main scripts, together with any custom, supplementary functions, should be uploaded to the Robotarium website.

---

[8]To create an account, potential users are asked to fill out an online form, providing their name, e-mail address, address, institute or employer, position or title, and reason(s) for using the Robotarium. The demographic and usage information is collected to help characterize and support the Robotarium userbase.

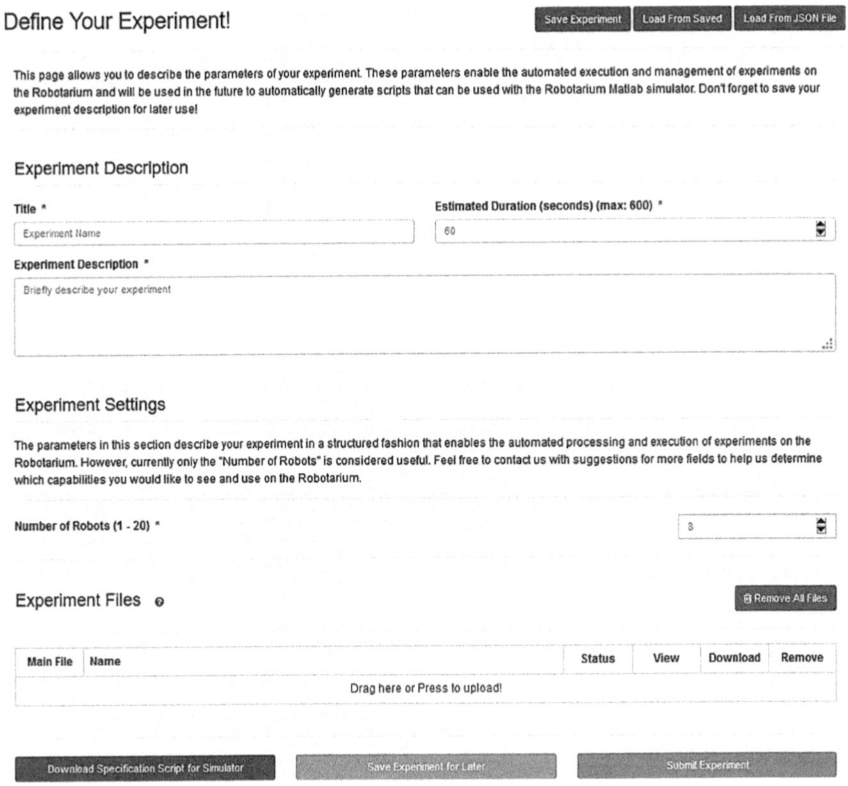

Define Your Experiment!

[Save Experiment] [Load From Saved] [Load From JSON File]

This page allows you to describe the parameters of your experiment. These parameters enable the automated execution and management of experiments on the Robotarium and will be used in the future to automatically generate scripts that can be used with the Robotarium Matlab simulator. Don't forget to save your experiment description for later use!

Experiment Description

Title *

Experiment Name

Estimated Duration (seconds) (max: 600) *

60

Experiment Description *

Briefly describe your experiment

Experiment Settings

The parameters in this section describe your experiment in a structured fashion that enables the automated processing and execution of experiments on the Robotarium. However, currently only the "Number of Robots" is considered useful. Feel free to contact us with suggestions for more fields to help us determine which capabilities you would like to see and use on the Robotarium.

Number of Robots (1 - 20) *

3

Experiment Files ⊘

[Remove All Files]

Main File	Name	Status	View	Download	Remove
Drag here or Press to upload!					

[Download Specification Script for Simulator] [Save Experiment for Later] [Submit Experiment]

Figure 9.14: The Robotarium experiment submission form.

When uploading the code through a drag-and-drop or file-searching interface, the users fill out an experiment description form containing a brief description of the experiment, an estimated experiment time, and the number of robots required, as shown in Figure 9.14.

The submitted user-code is stored in a database on a server located within the Robotarium facility. Users have the ability to delete their experiments from the database at any time through the web interface. This storage mechanism allows users to resubmit the same code at a later date through the web interface without local access to the files, exchange individual files of previous submissions, or request a Robotarium administrator's help with an error.[9]

---

[9] Submitted code is never accessed or viewed by a Robotarium team member without explicit user permission. Experiments are only accessed through the database autonomously by the server, and users with additional privacy concerns are able to upload non-human-readable files.

After the user submits their code to the website, it is pulled by the server and run in the simulation environment to check for problems, such as compilation errors, runtime errors, inclusion of libraries not available on the server, or excessive runtimes. If any issues are found during this verification step, the experiment is rejected and the user is returned a log file with specific details about the problem. If the simulation runs without error, the submission is forwarded to a first-in, first-out queue, waiting for its turn to run on the Robotarium.

The server automatically pulls the next verified experiment in the queue, and queries the robots on the testbed to select the N most charged robots, where N is the number of robots specified by the user through the submission form. If the recruited robot with the lowest voltage is not above a minimum voltage threshold, the system waits until enough robots are sufficiently charged. When the system is deemed ready to execute the experiment, the N robots selected depart from their charging stations to a random or user-specified initial configuration in the middle of the testbed, at which point the user-submitted script takes over.

After an experiment finishes executing, or is rejected as part of the verification step, the system autonomously sends an e-mail to the submitting user with information about the experiment's updated status. If an experiment runs successfully, the user is returned the overhead camera feed of the experiment, as well as any additional, requested data. As an example, Figure 9.15 shows a snapshot from a returned overhead video of an experiment (simulation as well as deployed on the real robots), that makes 20 GRITSBot X robots form the Georgia Institute of Technology logo.

9.3.2 The Robotarium Userbase

One advantage of the minimally invasive, constraint-based design methodology in Equation 9.10 is that it is *flexible*—the user-provided control input, u_{nom}, can be nearly anything—and, as such, provides broad support for the autonomy-on-demand concept. To stress this fact further, we here showcase a portfolio of experiments conducted on the Robotarium. In particular, we elucidate the geographic and topical reach of the platform, followed by a discussion about the types of experiments that have been conducted on the Robotarium.[10]

---

[10]The data presented here is taken over the three-year period, August 2017—July 2020, and is partially based on the findings reported in [431].

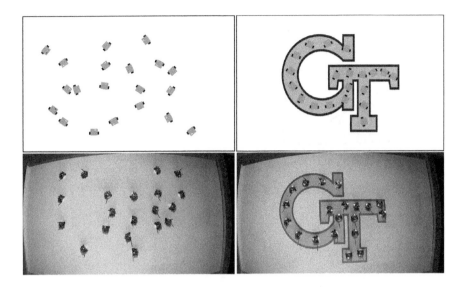

Figure 9.15: Screenshots from a simulation (Top), and returned video (Bottom), associated with a submission that drives 20 GRITSBot X robots to a formation in the shape of the Georgia Institute of Technology logo. Source: Fig. 4 in Wilson et al. (2020).

As the Robotarium is explicitly designed to democratize access to a high-fidelity, multi-robot testbed, having a sufficient geographic reach is an important attribute for this ambition to be realized. In fact, *the Robotarium has users from every continent except for Antarctica*, and, as such, has been quite effective at reaching a geographically diverse userbase. The corresponding geographic data is presented in Figure 9.16, where it can be seen that a majority of users come from North America, followed by a strong representation from Asia and Europe. South America, Oceania, and Africa are not equally well represented. But at least they are represented.

It is worth noticing that not all users who register for a Robotarium account end up submitting actual experiment scripts. In fact, only around 40% of the users who create an account end up deploying an experiment on the physical testbed. There are various reasons for this. Some users have explicitly stated that they only want to use the Robotarium simulation API, without the need for physical experiments. In addition, multiple accounts have been created by users working on team projects, with only one individual within the group uploading the code, which would account for a portion of the non-submitting users. There is also a typical lag of a few weeks between account

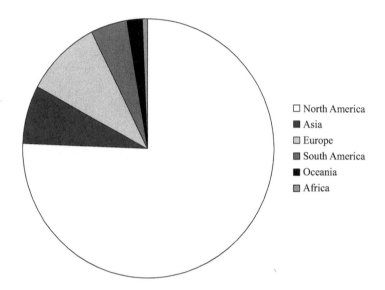

North America
Asia
Europe
South America
Oceania
Africa

Figure 9.16: The Robotarium user account demographics, grouped by self-reported continental regions. What is displayed are relative rather than absolute numbers, as the relative information tends to stay more or less constant over time.

creation and experiment submission, which means that the 40% number is probably somewhat lower than the actual number by a few percentage points. The data associated with the actual experiment submissions, broken down by geographical regions, is shown in Figure 9.17 where it can be seen that North America is even more over-represented in this category. Europe has overtaken Asia for second place, while Oceania and Africa are increasingly under-represented, as compared to the overall userbase [431].

Knowing *where* the Robotarium users are located, one can proceed to ask *why* they use the Robotarium. According to the responses to the questionnaire on the Robotarium website, the reasons vary quite significantly. The most common theme reported is, not surprisingly, related to a lack of local resources needed to establish and maintain a multi-robot testbed. This reason resonates directly with the ambition behind the Robotarium, i.e., to democratize access and lower the barrier to entry as many universities, research institutions, and secondary schools simply do not have robotic facilities. Those that do are typically not making them widely accessible to students and researchers outside of the labs that maintain them. One user from Brazil explained that the reason

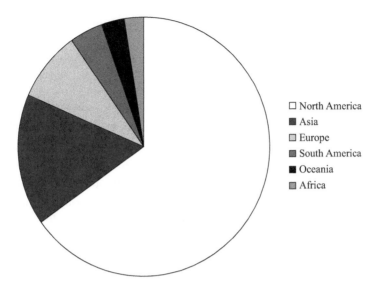

Figure 9.17: Users who have submitted actual experiments to be run on the Robotarium, organized by continental regions.

for using the Robotarium was to push beyond this barrier, and to appreciate what it takes to go from theory to application, stating:[11]

> *"The Robotarium will be used to learn principles of Robotics and Control and Automation Engineering. There is a lack of hardware at the university where I study. Therefore, practical lessons of those subjects are affected, as well as my learning experience. Robotarium would be an awesome solution to this specific matter."*

A less practical, but perhaps equally important reported usage reason was that actual robotics implementations have the potential to create excitement around research contributions. A Ph.D. candidate at a U.S. university used the Robotarium to test temporal logic-based control synthesis algorithms for multi-robot systems. In the experiment, a collection of 10 robots were asked to perform different tasks, such as populating or avoiding particular regions, visiting virtual charging stations, not visiting some regions until a specific event had occurred, etc., in order to mimic a complex emergency response scenario. When asked about the impact of the Robotarium, the Ph.D. candidate responded,

---

[11]The user-statements were originally reported in [431].

> *"The most important benefit of using the Robotarium for me is the motivation and inspiration I get from seeing my algorithm working on actual robots. Furthermore, my research attracts more attention when I use the Robotarium videos in my papers/presentations."*

Somewhat more surprising was the number of Robotarium users who used the Robotarium despite having local access to multi-robot testbeds. This usage category seems to have been driven by convenience more than anything else. It is, simply put, a tedious process to set up and deploy algorithms on a large team of robots. It is time consuming to program low-level collision-avoidance and motion controllers, to ensure that enough batteries are sufficiently charged, or to debug hardware complications and faulty communication protocols. The Robotarium sweeps these concerns to the side and gives users a chance to instead focus exclusively on algorithm design questions. As an example, another Ph.D. candidate used the Robotarium to test a distributed control strategy, generated by deep reinforcement learning, for fighting forest fires. When asked why the Robotarium was used, the answer was,

> *"The Robotarium allowed me to conduct hardware experiments without worrying about the overhead of setting up robots and other associated hardware first. The simulation environment allowed me to easily determine if the setup of a given experiment would show the details I needed, without also running the entire hardware system."*

Finally, some users choose to deploy their algorithms on the Robotarium because of its standardizing potential. Typically, educational modules or research experiments in multi-agent robotics are conducted on hardware infrastructure that is not easily replicable. As a result, produced code and results are hard to share, verify, and compare when robots, computational hardware, and tracking systems vary from lab to lab. Any code developed for an algorithm deployed on the Robotarium can be easily shared, modified, and built upon on the same hardware, and produce metrics that are comparable, without having to worry about the implementation specifics. A postdoctoral researcher at a U.S. research institution used the platform to test a multi-point rendezvous algorithm for this very reason, stating:

> *"[The Robotarium] has greatly enabled realistic and immediate verification of our multi-robot consensus and formation algorithms, through simulations and real experiments with the Robotarium platform. Using*

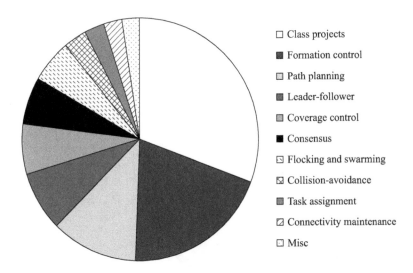

Figure 9.18: Types of experiments executed on the Robotarium, grouped into general, self-reported topics.

> *the platform, we were able to validate our algorithms and provide a reproducible code that can be implemented and tested anywhere."*

The user-statistics confirm that this remote-access testbed is having a global reach. The individual feedback from Robotarium users provides encouraging testimonials that go beyond geography in that the Robotarium is fulfilling its goal of democratizing access. And, almost as a side-effect, an effective, long-running and canonical autonomy-on-demand platform has been produced that can serve as a blueprint for how to structure other such facilities.

9.3.3 User Experiments

Geography aside, the real power of an autonomy-on-demand platform is only fully realized if it can be used to support a wide range of tasks. The Robotarium was originally intended to be primarily utilized by controls and robotics researchers focused on distributed systems who did not have the resources to test their multi-robot algorithms. As a result, the support tools developed, robots used, and recorded experimental data were chosen with this target audience in mind. Figure 9.18 depicts the types of experiments that have been executed on the Robotarium, grouped into general, topical research areas, based on the self-reported experiment descriptions. Indeed, the vast majority of these

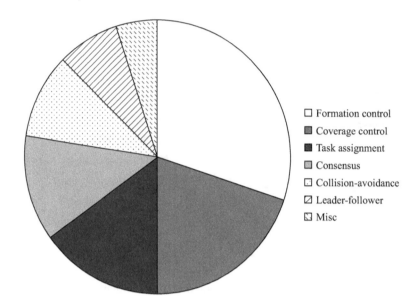

Figure 9.19: Peer-reviewed publications supported by the Robotarium, grouped according to research topic.

experiments are focused on distributed controls and robotics algorithms, i.e., the Robotarium structure is well-aligned with its intended use-cases.

Although the majority of research-oriented submissions to the Robotarium focus on "classical" multi-robot problems, such as formation and coverage control, the "Misc" category in Figure 9.18 hides some unexpected revelations. It turns out that among the experiments, one find submissions from myrmecologists and entomologists, social scientists, traffic engineers, and even ethicists, with topics including surprises such as queen selection mechanisms among ants, belief polarization effects in social networks, optimal traffic circle design, and implicit bias studies when using machine learning algorithms [431]. These use-cases are unintended but certainly welcomed consequences of building a flexible and remotely accessible autonomy-on-demand testbed.

Increasingly, the Robotarium experiments are being translated into scholarly products, such as peer-reviewed publications. In fact, research-based submissions have resulted in almost 100 published peer-reviewed papers during the Robotarium's first three years in operation [431]. Figure 9.19 shows the breakdown of these publications by topic. They give testament

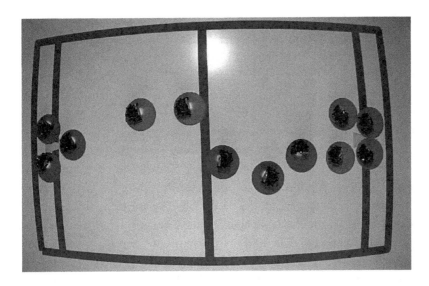

Figure 9.20: A snapshot from a project in a "Networked Control Systems" class, where robots are competing in a game of Capture the Flag. Source: Fig. 12 in Wilson et al. (2020).

to the fact that the experiments being executed on the Robotarium produce research-quality data that is enabling researchers to publish their algorithms, with the hardware validation happening at a remote facility.

Beyond research, the Robotarium platform is unexpectedly being leveraged by college-level students and educators for educational purposes. A significant number of users are developing project-based courses around the Robotarium; one project is depicted in Figure 9.20. For example, the Robotarium has been used to supplement a course in France called "Control of Multi-Agent Systems," where students implement consensus and formation control algorithms. In the United States, the Robotarium was used in a "Cyber-Physical Design and Analysis" course, offered as an online as well as an on-campus course, where online and on-campus students alike were tasked with developing Robotarium code to move robots between way-points by means of open-loop and closed-loop control, as a way of calling out the benefits of feedback.

9.3.4 Case Studies

To wrap up the discussion, we highlight a few specific submissions in order to elucidate how the submitted algorithms map directly onto the autonomy-on-demand framework.

Fault-Tolerant Rendezvous

One unfortunate consequence of distributed multi-robot algorithms is that they tend to be highly non-robust to malfunctioning or malicious robots, e.g., [150]. For example, the consensus protocol in Equation 2.6 will, if everything proceeds as planned, make the robots rendezvous at the centroid of the initial robot positions, e.g., [276, 316]. But, the consensus protocol is notoriously sensitive to outliers. If one robot malfunctions and simply stops moving, this immobile robot will completely dominate the overall behavior of the team in the sense that all other robots will move to the faulty robot's location. Or, even worse, a malicious robot could have infiltrated the team. And that robot could be moving towards some undesirable region. The consensus dynamics will consequently make all other robots follow the malicious robot, thereby letting themselves be hijacked [228, 326].

One of the first experiments executed on the Robotarium investigated this very issue, and introduced a notion of fault-tolerant rendezvous, based on algorithmic ideas from [325]. In this work, the robots achieve consensus by moving towards points within a safe, "non-faulty" set, while maintaining connectivity of the underlying disk-graph. Because this algorithm models robots as planar points with no spatial footprint, no native collision-avoidance accommodations were made by the algorithm. As such, the execution on the Robotarium utilizes the single integrator-to-unicycle mapping in Equation 9.17, as well as the safety barrier functions in Equation 9.4 in order to run effectively. Figure 9.21 (Left) shows the results of this experiment.

Passivity-Based Attitude Synchronization

Attitude synchronization is highly related to Reynolds' alignment concept [356] and is sometimes referred to as flocking, e.g., [318], in that the objective is to make all robots move in the same general direction. Adhering to the notation in Equation 9.12, what is required is that $\lim_{t\to\infty} \phi_i(t) - \phi_j(t) = 0$, $\forall i, j \in [N]$. The reason why this objective cannot simply be achieved by running the consensus equation over the robot orientations, rather than their positions, is that orientations are angles, i.e., they belong to S^1 rather than to $\mathbb{R}$. And all sorts of strange things happen if one ignores the fact that θ and $\theta + 2\pi k$, $k \in \mathbb{Z}$, correspond to the same direction.

To illustrate this point, the attitude synchronization algorithm from [193] was implemented on the Robotarium. Utilizing the passivity property of general rigid body motions, the deployed algorithm was designed to

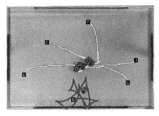 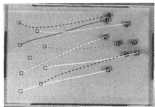 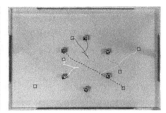

Figure 9.21: User experiments on the Robotarium that explicitly call out the inherent flexibility and safety mechanics associated with the autonomy-on-demand framework. Left: Robots are executing a fault-tolerant rendezvous algorithm that allows them to ignore faulty or malicious robots. Middle: Attitude synchronization, which ensures that the robots end up moving in the same general direction, is pursued using only locally available information. Right: A desired geometric shape is assembled based on a sparse set of specified inter-robot relationships.

achieve attitude synchronization for a group of rigid bodies, with only local information exchanges. The execution of this algorithm in the Robotarium relies on the ability to specify an information exchange graph, and to map from single integrator to unicycle dynamics, which are, as we have seen, two attributes that are readily available in the Robotarium. Figure 9.21 (Middle) displays the outcome of this experiment.

Distributed Formation Control
Typically, enough inter-robot distances must be specified in order to uniquely determine the shape of a formation, up to arbitrary rotations and translations. This requirement is captured by the graph rigidity concept, e.g., [15, 134, 383]. However, if relative orientations are added to the mix—in addition to distances—one can potentially get by with fewer specified inter-robot relationships [67, 117].

The idea of using a sparse set of specifications was called out by an experiment on the Robotarium that pursued a distributed formation control algorithm, based on the developments in [138, 139]. The controllers use relative position measurements in local coordinate frames, and side-step the need for any explicit inter-robot communication. The algorithm moreover assumes that the robots are points in the plane, and does not consider collision-avoidance. The successful execution on the Robotarium therefore depends, once again, on the safety functionality provided by the autonomy-on-demand formalism. The outcome of this experiment is shown in Figure 9.21 (Right).

These three case studies constitute a small sample of all the different experiments, or missions, that the Robotarium robots have participated in during their first three years. To get a better feeling for the scope of these missions, we here catalog other common types of experiments supported by the Robotarium from an application-centric vantage point. This catalog also serves as a snapshot in time, showcasing particular directions being pursued by the multi-agent robotics research community.

Autonomous Vehicles

A number of scenarios pertaining to the use of autonomous vehicles in complex traffic environments have been considered by the Robotarium userbase, including:

- Lane-following behaviors for self-driving cars that avoid the types of oscillatory effects known to cause traffic congestion;
- Mixed human/autonomous vehicle traffic, including rules that let the autonomous vehicles smooth out traffic patterns;
- On-demand delivery systems, such as self-driving cars or unmanned ground vehicles, used for time-critical deliveries, e.g., for delivering perishables in a particular time window;
- Traffic circle layout and rule design for optimizing traffic throughput; and
- Closed-loop traffic signal control that minimizes congestion and alleviates the negative environmental effects associated with stop-go traffic.

Social Insects

The Robotarium has seen its arena turned into ant and termite colonies, bee hives, and even geometric moth pillars, with the investigations covering issues such as:

- Queen selection mechanisms, whereby ants interact competitively in a pairwise manner to elect new queens;
- Pheromone-based foraging strategies that tag the already visited parts of an environment for the purpose of achieving more effective coverage;
- Formation control algorithms for achieving the types of shapes observed in nature; and

- Collaborative manipulation, involving multiple robots teaming up to transport objects that are heavier or bulkier than what any individual robot could manipulate, akin to the way ants collaborate when transporting food back to the mound.

Urban Navigation

Some of the more common environments displayed on the Robotarium arena by the overhead projector are various urban landscapes. Buildings provide line-of-sight obstructions, render wireless commutations unreliable, and provide obstacle-rich vistas. These complications have direct bearing on the control design, and topics investigated in this context include:

- Search-and-rescue applications, where teams of robots must search an area in order to locate potential targets, e.g., victims after natural disasters such as earthquakes, in order to report their locations back to base stations;
- Perimeter defense, concerning the problem of using patrolling robots to encircle an area so as to minimize intruders' ability to slip through the protected perimeter; and
- Communication-denied or intermittent coordination in urban canyons, arising from the need to ensure that robot teams are able to perform tasks in a coordinated manner, despite the lack of reliable communication channels.

Precision Agriculture

Farm fields and fruit orchards are increasingly serving as hosts to autonomous vehicles. This trend is also reflected in the Robotarium experiment catalog, which include:

- Persistent monitoring scenarios, where robots are present among the plants for sustained periods of time for the purpose of tending to individual plants in terms of meeting their water, pesticide, and fertilizer needs;
- Targeted delivery of water or nutrients using aerial delivery vehicles, which requires solutions to taveling salesperson problems, with or without temporal constraints; and
- Optimal sweep patterns for planting or harvesting, which, for example, are needed when moving autonomous tractors across fields in a coordinated manner.

Warehousing and Manufacturing

Robots have historically been used extensively in manufacturing settings and they are increasingly performing fetch-and-carry tasks in warehouses as well. Such deployments typically involve carefully choreographed maneuvers that are effective, but not particularly flexible. As more customization and small-batch production is coming online and robots are operating more and more alongside human workers, a number of new issues must be addressed in this classical robotics domain, such as:

- Task assignment problems that arise when multiple robots can be called upon to carry out any one of a number of different tasks;
- Path-planning algorithms for sending multiple robots across a cluttered warehouse workspace in an effective and orderly manner;
- Area coverage problems that present themselves naturally in cleaning applications, such as when a robot, or team of robots, are to vacuum a manufacturing floor;
- Simultaneous localization and mapping problems that involve the issue of having robots keep track of where in the environment they are, as well as map the layout of the area in which they are deployed. This is a crucial capability in many indoor robotics applications; and
- Human-robot interactions for ensuring that robots can operate safely, effectively, and in a transparent and non-threatening manner alongside human workers, such as during the assembly phase of a car or air-craft manufacturing process, or when boxes should be jointly packed by humans and robots in delivery applications.

Natural Ecosystems

The natural world is typically messy, dynamic, and not overly well-mapped, which calls for a different set of algorithms and considerations than what is required in carefully curated and engineered environments. This issue has been investigated in the Robotarium through a number of user-submitted experiments, including:

- Containment of evolving fronts, e.g., forest fires or oil spills, that must be tracked, contained, and possibly even cleaned up by a team of robots;
- Navigation through cluttered environments, which typically requires a different set of path-planning algorithms than what can be used for human-made environments. Rather than having a space be either free or occupied, it is much more nuanced and fluid out in the natural world,

where certain areas are more or less easy/hard for the robots to navigate; and

- Learning to engage safely with objects, whereby robots not only have to navigate unknown terrains, but also learn which types of terrains and object-types can be safely traversed, and which should be avoided at all costs.

Interplanetary Exploration

Robots are particularly well-suited for environments that are hard to reach by humans, such as other planets. To this end, the Robotarium has been turned into lunar and Martian landscapes, in order to promote inquiries in areas such as:

- Coverage, where a team of robots are to spread out across an unknown domain for the purpose of detecting interesting or potentially harmful phenomena and events;
- Exploration, which is the dynamic counterpart to coverage, where the robots are not only asked to effectively search for key features in the environment, but they have to do so dynamically, rather than establish static configurations; and
- Foraging, whereby the robots are to detect objects of interest in the environment, and return these objects to select base locations for the purpose of further analysis.

The last two topical areas, *Natural Ecosystems* and *Interplanetary Exploration*, are particularly pleasing in that their deployments on the Robotarium take the *robot ecology* story full circle. What initially started with a mood picture from Mars, where the Mars rovers would exhibit remarkable longevity thanks to a low-energy lifestyle, a tight coupling between rovers and their Martian environment, and a focus on robot survival, ended up connecting long-duration autonomy to ecology, by observing that *the tight coupling between robots and their habitats is key to understanding how robots can be deployed in natural environments over truly long time-scales.*

Bibliography

[1] R. R. Abraham, J. E. Marsden, and T. Ratiu, *Manifolds, Tensor Analysis, and Applications*. Springer Science & Business Media, Berlin/Heidelberg, Germany, 2012.

[2] P. A. Abrams, "The evolution of predator-prey interactions: Theory and evidence," *Annual Review of Ecology and Systematics*, vol. 31, no. 1, pp. 79–105, 2000.

[3] C. Adjih, E. Baccelli, E. Fleury, G. Harter, N. Mitton, T. Noel, R. Pissard-Gibollet, F. Saint-Marcel, G. Schreiner, J. Vandaele, and T. Watteyne, "FIT IoT-LAB: A large scale open experimental IoT testbed," *IEEE World Forum on Internet of Things*, pp. 459–464, Dec. 2015.

[4] J. Aguilar, D. Monaenkova, V. Linevich, W. Savoie, B. Dutta, H. S. Kuan, M. D. Betterton, M. A. D. Goodisman, and D. I. Goldman, "Collective clog control: Optimizing traffic flow in confined biological and robophysical excavation," *Science*, vol. 361, no. 6403, pp. 672–677, 2018.

[5] A. K. Akametalu, J. F. Fisac, J. H. Gillula, S. Kaynama, M. N. Zeilinger, and C. J. Tomlin, "Reachability-based safe learning with Gaussian processes," *IEEE Conference on Decision and Control*, pp. 1424–1431, Dec. 2014.

[6] Z. Ákos, M. Nagy, S. Leven, and T. Vicsek, "Thermal soaring flight of birds and unmanned aerial vehicles," *Bioinspiration & Biomimetics*, vol. 5, no. 4, p. 045003, 2010.

[7] J. Alcock and D. R. Rubenstein, *Animal Behavior: An Evolutionary Approach*. Sinauer Associates, Sunderland, MA, 2009.

[8] F. Allgöwer and A. Zheng, *Nonlinear Model Predictive Control*. Birkhäuser, Basel, Switzerland, 2012.

[9] J. Alonso-Mora, A. Breitenmoser, M. Rufli, P. Beardsley, and R. Siegwart, "Optimal reciprocal collision avoidance for multiple non-holonomic robots," *Distributed Autonomous Robotic Systems*, pp. 203–216, 2013.

[10] E. Alpaydin, *Introduction to Machine Learning*. MIT Press, Cambridge, MA, 2020.

[11] A. D. Ames, S. Coogan, M. Egerstedt, G. Notomista, K. Sreenath, and P. Tabuada, "Control barrier functions: Theory and applications," *European Control Conference*, 2019.

[12] A. D. Ames, K. Galloway, K. Sreenath, and J. W. Grizzle, "Rapidly exponentially stabilizing control Lyapunov functions and hybrid zero dynamics," *IEEE Transactions on Automatic Control*, vol. 59, no. 4, pp. 876–891, 2014.

[13] A. D. Ames, J. W. Grizzle, and P. Tabuada, "Control barrier function based quadratic programs with application to adaptive cruise control," *IEEE Conference on Decision and Control*, pp. 6271–6278, 2014.

[14] A. D. Ames, X. Xu, J. W. Grizzle, and P. Tabuada, "Control barrier function based quadratic programs for safety critical systems," *IEEE Transactions on Automatic Control*, vol. 62, no. 8, pp. 3861–3876, 2017.

[15] B. D. O. Anderson, C. Yu, B. Fidan, and J. M. Hendrickx, "Rigid graph control architectures for autonomous formations," *IEEE Control Systems Magazine*, vol. 28, no. 6, pp. 48–63, 2008.

[16] T. L. Anderson and M. Donath, "Animal behavior as a paradigm for developing robot autonomy," *Robotics and Autonomous Systems*, vol. 6, no. 1–2, pp. 145–168, 1990.

[17] H. Ando, Y. Oasa, I. Suzuki, and M. Yamashita, "Distributed memoryless point convergence algorithm for mobile robots with limited

visibility," *IEEE Transactions on Robotics and Automation*, vol. 15, no. 5, pp. 818–828, 1999.

[18] S. Aoshima, T. Tsujimura, and T. Yabuta, "A wire mobile robot with multi-unit structure," *IEEE/RSJ International Conference on Intelligent Robots and Systems*, pp. 414–421, Sep. 1989.

[19] S. Appanah and H. T. Chan, "Thrips: The pollinators of some dipterocarps," *Malaysian Forester*, vol. 44, no. 2–3, pp. 234–252, 1981.

[20] M. A. Arbib, "Schema theory," *The Encyclopedia of Artificial Intelligence*, vol. 2, pp. 1427–1443. John Wiley & Sons, New York, NY, 1992.

[21] M. A. Arbib, *Brain Theory and Neural Networks*. MIT Press, Cambridge, MA, 1995.

[22] R. C. Arkin, "Motor schema-based mobile robot navigation," *International Journal of Robotics Research*, vol. 8, no. 4, pp. 92–112, 1989.

[23] R. C. Arkin and M. Egerstedt, "Temporal heterogeneity and the value of slowness in robotic systems," *IEEE International Conference on Robotics and Biomimetics*, pp. 1000–1005, 2015.

[24] R. C. Arkin, "Behavior-based robot navigation for extended domains," *Adaptive Behavior*, vol. 1, no. 2, pp. 201–225, 1992.

[25] R. C. Arkin, *Behavior-Based Robotics*. MIT Press, Cambridge, MA, 1998.

[26] K. Arrow, B. Bolin, R. Costanza, P. Dasgupta, C. Folke, C. S. Holling, B. O. Jansson, S. Levin, K. Mäler, C. Perrings, and D. Pimentel, "Economic growth, carrying capacity, and the environment," *Ecological Economics*, vol. 15, no. 2, pp. 91–95, 1995.

[27] I. Asimov, "Runaround," in *I, Robot*. Bantam Dell, New York, NY, 1950.

[28] A. Aswani, H. Gonzalez, S. S. Sastry, and C. J. Tomlin, "Provably safe and robust learning- based model predictive control," *Automatica*, vol. 49, no. 5, pp. 1216–1226, 2013.

[29] J. P. Aubin, *Viability Theory*. Springer Science & Business Media, Berlin/Heidelberg, Germany, 2009.

[30] A. Bacciotti and F. Ceragioli, "Stability and stabilization of discontinuous systems and nonsmooth Lyapunov functions," *ESAIM: Control, Optimisation and Calculus of Variations*, vol. 4, pp. 361–376, 1999.

[31] A. Bacciotti and F. Ceragioli, "Nonpathological Lyapunov functions and discontinuous Carathéodory systems," *Automatica*, vol. 42, no. 3, pp. 453–458, Mar. 2006.

[32] I. L. Bajec and F. H. Heppner, "Organized flight in birds," *Animal Behaviour*, vol. 78, no. 4, pp. 777–789, 2009.

[33] P. Bakker and Y. Kuniyoshi, "Robot see, robot do: An overview of robot imitation," *AISB96 Workshop on Learning in Robots and Animals*, pp. 3–11, 1996.

[34] E. Bakolas and P. Tsiotras, "The Zermelo-Voronoi diagram: A dynamic partition problem," *Automatica*, vol. 46, pp. 2059–2067, 2010.

[35] T. Balch, "Hierarchic social entropy: An information theoretic measure of robot group diversity," *Autonomous Robots*, vol. 8, no. 3, pp. 209–238, 2000.

[36] T. Balch and R. C. Arkin, "Behavior-based formation control for multi-robot teams," *IEEE Transactions on Robotics and Automation*, vol. 14, no. 6, pp. 926–939, 1998.

[37] T. Balch, F. Dellaert, A. Feldman, A. Guillory, C. L. Isbell, Z. Khan, S. C. Pratt, A. N. Stein, and H. Wilde, "How multirobot systems research will accelerate our understanding of social animal behavior," *Proceedings of the IEEE*, vol. 94, no. 7, pp. 1445–1463, 2006.

[38] D. Ball, P. Ross, A. English, P. Milani, D. Richards, A. Bate, B. Upcroft, G. Wyeth, and P. Corke, "Farm workers of the future: Vision-based robotics for broad-acre agriculture," *IEEE Robotics & Automation Magazine*, vol. 24, no. 3, pp. 97–107, 2017.

[39] R. Beckers, S. Goss, J. L. Deneubourg, and J. M. Pasteels, "Colony size, communication and ant foraging strategy," *Psyche*, vol. 96, no. 3–4, pp. 239–256, 1989.

[40] R. Beckers, O. E. Holland, and J. L. Deneubourg, "From local actions to global tasks: Stigmergy and collective robotics," *Prerational Intelligence: Adaptive Behavior and Intelligent Systems Without Symbols and Logic*, vol. 1, pp. 1008–1022, 2000.

[41] R. D. Beer, H. J. Chiel, and L. S. Sterling, "A biological perspective on autonomous agent design," *Robotics and Autonomous Systems*, vol. 6, no. 1–2, pp. 169–186, 1990.

[42] J. G. Bellingham and K. Rajan, "Robotics in remote and hostile environments," *Science*, vol. 318, no. 5853, pp. 1098–1102, 2007.

[43] M. Bennewitz, W. Burgard, and S. Thrun, "Optimizing schedules for prioritized path planning of multi-robot systems," *IEEE International Conference on Robotics and Automation*, pp. 271–276, 2001.

[44] T. C. Bergstrom, "Evolution of social behavior: Individual and group selection," *Journal of Economic Perspectives*, vol. 16, no. 2, pp. 67–88, 2002.

[45] F. Berkenkamp, R. Moriconi, A. P. Schoellig, and A. Krause, "Safe learning of regions of attraction for uncertain, nonlinear systems with Gaussian processes," *IEEE Conference on Decision and Control*, pp. 4661–4666, Dec. 2016.

[46] S. Berman, Q. Lindsey, M. S. Sakar, V. Kumar, and S. C. Pratt, "Experimental study and modeling of group retrieval in ants as an approach to collective transport in swarm robotic systems," *Proceedings of the IEEE*, vol. 99, no. 9, pp. 1470–1481, 2011.

[47] D. P. Bertsekas and J. N. Tsitsiklis, *Parallel and Distributed Computation*. Prentice-Hall, Upper Saddle River, NJ, 1989.

[48] B. Bethke, J. How, and J. Vian, "Multi-UAV persistent surveillance with communication constraints and health management," *AIAA Guidance, Navigation, and Control Conference*, p. 5654, 2009.

[49] S. P. Bhat and D. S. Bernstein, "Finite-time stability of continuous autonomous systems," *SIAM Journal on Control and Optimization*, vol. 38, no. 3, pp. 751–766, 2000.

[50] J. Biesiadecki, C. Leger, and M. W. Maimone, "Tradeoffs between directed and autonomous driving on the Mars exploration rovers," *International Journal of Robotics Research*, vol. 26, no. 1, pp. 91–104, Jan. 2007.

[51] J. Billingsley, A. Visala, and M. Dunn, "Robotics in agriculture and forestry," in *Springer Handbook of Robotics*, edited by B. Siciliano and O. Khatib, pp. 1065–1077. Springer-Verlag, Berlin/Heidelberg, Germany, 2008.

[52] F. Blanchini, "Set invariance in control," *Automatica*, vol. 35, no. 11, pp. 1747–1767, 1999.

[53] I. M. Bomze, "Lotka-Volterra equation and replicator dynamics: A two-dimensional classification," *Biological Cybernetics*, vol. 48, no. 3, pp. 201–211, 1983.

[54] I. M. Bomze, "Lotka-Volterra equation and replicator dynamics: New issues in classification," *Biological Cybernetics*, vol. 72, no. 5, pp. 447–453, 1995.

[55] J. M. Bony, "Principe du maximum, inégalité de harnack et unicité du probleme de cauchy pour les opérateurs elliptiques dégénérés," *Annals of the Institute Fourier*, vol. 19, no. 1, pp. 277–304, 1969.

[56] U. Borrmann, L. Wang, A. D. Ames, and M. Egerstedt, "Control barrier certificates for safe swarm behavior," *IFAC Conference on Analysis and Design of Hybrid Systems*, Oct. 2015.

[57] H. O. Box, *Organisation in Animal Communities*. Butterworths, Oxford, UK, 1973.

[58] S. Boyd and L. Vandenberghe, *Convex Optimization*. Cambridge University Press, Cambridge, UK, 2004.

[59] V. Braitenberg, *Vehicles: Experiments in Synthetic Psychology*. MIT Press, Cambridge, MA, 1986.

[60] H. Brezis, "On a characterization of flow-invariant sets," *Communications on Pure and Applied Mathematics*, vol. 23, no. 2, pp. 261–263, 1970.

[61] J. L. Bronstein, *Mutualism*. Oxford University Press, Oxford, UK, 2015.

[62] R. A. Brooks, "A robust layered control system for a mobile robot," *IEEE Journal on Robotics and Automation*, vol. 2, no. 1, pp. 14–23, 1986.

[63] R. A. Brooks, "Elephants don't play chess," *Robotics and Autonomous Systems*, vol. 6, no. 1–2, pp. 3–15, 1990.

[64] R. A. Brooks, "Intelligence without representation," *Artificial Intelligence*, vol. 47, no. 1–3, pp. 139–159, 1991.

[65] B. Brumitt, A. Stentz, and M. Hebert, "Autonomous driving with concurrent goals and multiple vehicles: Mission planning and architecture," *Autonomous Robots*, vol. 11, no. 2, pp. 103–115, 2001.

[66] A. E. Bryson, *Applied Optimal Control: Optimization, Estimation and Control*. Routledge, Abingdon, UK, 2018.

[67] I. Buckley and M. Egerstedt, "Infinitesimally shape-similar motions using relative angle measurements," *IEEE/RSJ International Conference on Intelligent Robots and Systems*, pp. 1077–1082, 2017.

[68] F. Bullo, J. Cortés, and S. Martinez, *Distributed Control of Robotic Networks: A Mathematical Approach to Motion Coordination Algorithms*. Princeton University Press, Princeton, NJ, 2009.

[69] W. Burgard, M. Moors, D. Fox, R. Simmons, and S. Thrun, "Collaborative multi-robot exploration," *IEEE International Conference on Robotics and Automation*, vol. 1, pp. 476–481, 2000.

[70] S. Camazine, J. L. Deneubourg, N. R. Franks, J. Sneyd, E. Bonabeau, and G. Theraula, *Self-Organization in Biological Systems*. Princeton University Press, Princeton, NJ, 2003.

[71] M. Campbell, M. Egerstedt, J. P. How, and R. M. Murray, "Autonomous driving in urban environments: Approaches, lessons and challenges," *Philosophical Transactions of the Royal Society of London A: Mathematical, Physical and Engineering Sciences*, vol. 368, no. 1928, pp. 4649–4672, 2010.

[72] J. F. Canny, *The Complexity of Robot Motion Planning*. MIT Press, Cambridge, MA, 1988.

[73] G. M. Carr and D. W. Macdonald, "The sociality of solitary foragers: A model based on resource dispersion," *Animal Behaviour*, vol. 34, no. 5, pp. 1540–1549, 1986.

[74] J. Carsten, A. Rankin, D. Ferguson, and A. Stentz, "Global path planning on board the Mars exploration rovers," *IEEE Aerospace Conference*, Mar. 2007.

[75] J. Carsten, A. Rankin, D. Ferguson, and A. Stentz, "Global planning on the Mars exploration rovers: Software integration and surface testing," *Journal of Field Robotics*, vol. 26, no. 4, pp. 337–357, Apr. 2009.

[76] G. A. Casan, E. Cervera, A. A. Moughlbay, J. Alemany, and P. Martinet, "ROS-based online robot programming for remote education and training," *IEEE International Conference on Robotics and Automation*, pp. 6101–6106, 2015.

[77] E. Castello, T. Yamamoto, L. Dalla, W. Liu, A. F. T. Winfield, Y. Nakamura, and H. Ishiguro, "Adaptive foraging for simulated and real robotic swarms: The dynamical response threshold approach," *Swarm Intelligence*, vol. 10, no. 1, pp. 1–31, 2016.

[78] F. Celi, L. Wang, L. Pallottino, and M. Egerstedt, "Deconfliction of motion paths with traffic inspired rules," *IEEE Robotics and Automation Letters*, vol. 4, no. 2, pp. 2227–2234, 2019.

[79] E. L. Charnov, "Optimal foraging: Attack strategy of a mantid," *The American Naturalist*, vol. 110, no. 971, pp. 141–151, 1976.

[80] E. L. Charnov, "Optimal foraging: The marginal value theorem," University of New Mexico Digital Repository, Albuquerque, NM, 1976.

[81] J. M. Chase, "Ecological niche theory," in *The Theory of Ecology*, edited by S. Scheiner and M. R. Willig, pp. 93–107. University of Chicago Press, Chicago, IL, 2011.

[82] K. H. Cho, Y. H. Jin, H. M. Kim, H. Moon, J. C. Koo, and H. R. Choi, "Caterpillar-based cable climbing robot for inspection of suspension bridge hanger rope," *IEEE International Conference on Automation Science and Engineering*, pp. 1059–1062, 2013.

[83] H. M. Choset, S. Hutchinson, K. M. Lynch, G. Kantor, W. Burgard, L. E. Kavraki, and S. Thrun, *Principles of Robot Motion: Theory, Algorithms, and Implementation*. MIT Press, Cambridge, MA, 2005.

[84] J. Clark and R. Fierro, "Cooperative hybrid control of robotic sensors for perimeter detection and tracking," *American Control Conference*, pp. 3500–3505, 2005.

[85] F. Clarke, *Optimization and Nonsmooth Analysis*. Society for Industrial and Applied Mathematics, Philadelphia, PA, 1990.

[86] B. F. Clough and K. Scott, "Allometric relationships for estimating above-ground biomass in six mangrove species," *Forest Ecology and Management*, vol. 27, no. 2, pp. 117–127, 1989.

[87] R. C. Connor, M. R. Heithaus, and L. M. Barre, "Superalliance of bottlenose dolphins," *Nature*, vol. 397, no. 6720, pp. 571–572, 1999.

[88] R. C. Connor, R. A. Smolker, and A. F. Richards, "Two levels of alliance formation among male bottlenose dolphins (*Tursiops* sp.)," *Proceedings of the National Academy of Sciences*, vol. 89, no. 3, pp. 987–990, 1992.

[89] R. C. Connor, R. Smolker, and L. Bejder, "Synchrony, social behaviour and alliance affiliation in Indian Ocean bottlenose dolphins, *Tursiops aduncus*," *Animal Behaviour*, vol. 72, no. 6, pp. 1371–1378, 2006.

[90] R. C. Connor, M. R. Heithaus, and L. M. Barre, "Complex social structure, alliance stability and mating access in a bottlenose dolphin 'super-alliance'," *Proceedings of the Royal Society of London. Series B: Biological Sciences*, vol. 268, no. 1464, pp. 263–267, 2001.

[91] S. M. Cooper, "Optimal hunting group size: The need for lions to defend their kills against loss to spotted hyaenas," *African Journal of Ecology*, vol. 29, no. 2, pp. 130–136, 1991.

[92] J. Cortés, "Discontinuous dynamical systems," *IEEE Control Systems Magazine*, vol. 28, no. 3, pp. 36–73, Jun. 2008.

[93] J. Cortés and M. Egerstedt, "Coordinated control of multi-robot systems: A survey," *SICE Journal of Control, Measurement, and System Integration*, vol. 10, no. 6, pp. 495–503, 2017.

[94] J. Cortés, S. Martinez, and F. Bullo, "Spatially-distributed coverage optimization and control with limited-range interactions," *ESAIM: Control, Optimisation and Calculus of Variations*, vol. 11, no. 4, pp. 691–719, 2005.

[95] J. Cortés, S. Martinez, T. Karatas, and F. Bullo, "Coverage control for mobile sensing networks," *IEEE Transactions on Robotics and Automation*, vol. 20, no. 2, pp. 243–255, 2004.

[96] J. Cortés, S. Martinez, and F. Bullo, "Robust rendezvous for mobile autonomous agents via proximity graphs in arbitrary dimensions," *IEEE Transactions on Automatic Control*, vol. 51, no. 8, pp. 1289–1298, 2006.

[97] I. D. Couzin, J. Krause, N. R. Franks, and S. A. Levin, "Effective leadership and decision-making in animal groups on the move," *Nature*, vol. 433, no. 7025, pp. 513–516, 2005.

[98] I. D. Couzin, J. Krause, R. James, G. Ruxton, and N. Franks, "Collective memory and spatial sorting in animal groups," *Journal of Theoretical Biology*, vol. 218, pp. 1–11, 2002.

[99] J. P. de la Croix and M. Egerstedt, "Group-size selection for a parameterized class of predator-prey models," *Mathematical Theory of Networks and Systems*, Groningen, The Netherlands, July 2014.

[100] A. C. Crombie, "The effect of crowding upon the oviposition of grain-infesting insects," *Journal of Experimental Biology*, vol. 19, no. 3, pp. 311–340, 1942.

[101] E. Curio, *The Ethology of Predation*. Springer Science & Business Media, Berlin/Heidelberg, Germany, 2012.

[102] E. Danchin, L. E. Girladeau, and F. Cezilly, *Behavioural Ecology: An Evolutionary Perspective on Behaviour*. Oxford University Press, Oxford, UK, 2008.

[103] R. D'Andrea and G. E. Dullerud, "Distributed control design for spatially interconnected systems," *IEEE Transactions on Automatic Control*, vol. 49, no. 9, pp. 1478–1495, 2003.

[104] C. Daniel and J. Besenhard, *Handbook of Battery Materials*. Wiley-VCH Verlag, Weinheim, Germany, 2012.

[105] C. Darwin, *The Origin of Species*. P. F. Collier & Son, New York, NY, 1909.

[106] J. Davenport, *Animal Life at Low Temperature*. Springer Science & Business Media, Berlin/Heidelberg, Germany, 2012.

[107] N. B. Davies, J. R. Krebs, and S. A. West, *An Introduction to Behavioural Ecology*. John Wiley & Sons, Hoboken, NJ, 2012.

[108] M. L. Davis and D. A. Cornwell, *Introduction to Environmental Engineering*. McGraw-Hill, New York, NY, 2008.

[109] R. Dawkins, *The Extended Phenotype*. Oxford University Press, Oxford, UK, 1982, vol. 8.

[110] B. Dawson, "The impact of technology insertions on organizations," *Human Factors Integration Defense Technology Centre*, University of Southampton, Southampton, UK, 2007.

[111] P. Dayawansa and C. F. Martin, "A converse Lyapunov theorem for a class of dynamical systems which undergo switching," *IEEE Transactions on Automatic Control*, vol. 44, no. 4, pp. 751–760, 1999.

[112] P. Debenest, M. Guarnieri, K. Takita, E. F. Fukushima, S. Hirose, K. Tamura, A. Kimura, H. Kubokawa, N. Iwama, and F. Shiga, "Expliner—Robot for inspection of transmission lines," *IEEE International Conference on Robotics and Automation*, pp. 3978–3984, May 2008.

[113] M. P. Deisenroth, D. Fox, and C. E. Rasmussen, "Gaussian processes for data-efficient learning in robotics and control," *IEEE Transactions on Pattern Analysis and Machine Intelligence*, vol. 37, no. 2, pp. 408–423, 2013.

[114] F. Dellaert and M. Kaess, "Square Root SAM: Simultaneous localization and mapping via square root information smoothing," *International Journal of Robotics Research*, vol. 25, no. 12, pp. 1181–1203, 2006.

[115] J. L. Deneubourg and S. Goss, "Collective patterns and decision-making," *Ethology Ecology & Evolution*, vol. 1, no. 4, pp. 295–311, 1989.

[116] J. Derenick, N. Michael, and V. Kumar, "Energy-aware coverage control with docking for robot teams," *IEEE/RSJ International Conference on Intelligent Robots and Systems*, pp. 3667–3672, 2011.

[117] J. P. Desai, J. Ostrowski, and V. Kumar, "Controlling formations of multiple mobile robots," *IEEE International Conference on Robotics and Automation*, vol. 4, pp. 2864–2869, 1998.

[118] Y. Diaz-Mercado, S. Lee, and M. Egerstedt, "Human-swarm interactions via coverage of time-varying densities," in *Trends in Control and Decision-Making for Human–Robot Collaboration Systems*, edited by Y. Wang and F. Zhang, pp. 357–385. Springer International Publishing, Cham, Switzerland, 2017.

[119] J. R. M. de Dios, A. Jimenez-Gonzalez, A. de San Bernabe, and A. Ollero, *A Remote Integrated Testbed for Cooperating Objects*. Springer Science & Business Media, Berlin/Heidelberg, Germany, 2013.

[120] J. C. Doyle, "Guaranteed margins for LQG regulators," *IEEE Transactions on Automatic Control*, vol. 23, no. 4, pp. 756–757, 1978.

[121] Q. Du, M. Emelianenko, and L. Ju, "Convergence of the Lloyd algorithm for computing centroidal Voronoi tessellations," *SIAM Journal on Numerical Analysis*, vol. 44, no. 1, pp. 102–119, 2006.

[122] Q. Du, V. Faber, and M. Gunzburger, "Centroidal voronoi tessellations: Applications and algorithms," *SIAM Review*, vol. 41, no. 4, pp. 637–676, 1999.

[123] Q. Du and D. Wang, "The optimal centroidal Voronoi tessellations and the Gersho's conjecture in the three-dimensional space," *Computers & Mathematics with Applications*, vol. 49, no. 9–10, pp. 1355–1373, 2005.

[124] M. Dunbabin and L. Marques, "Robots for environmental monitoring: Significant advancements and applications," *IEEE Robotics & Automation Magazine*, vol. 19, no. 1, pp. 24–39, 2012.

[125] A. Dussutour, V. Fourcassie, D. Helbing, and J. L. Deneubourg, "Optimal traffic organization in ants under crowded conditions," *Nature*, vol. 428, no. 6978, p. 70, 2004.

[126] M. Egerstedt and M. Govindarasu, *Accessible Remote Testbeds: Opportunities, Challenges, and Lessons Learned*. US National Science Foundation Workshop Report, Nov. 2015.

[127] M. Egerstedt and X. Hu, "Formation constrained multi-agent control," *IEEE Transactions on Robotics and Automation*, vol. 17, no. 6, pp. 947–951, 2001.

[128] M. Egerstedt, J. N. Pauli, S. Hutchinson, and G. Notomista, "Robot ecology: Constraint-based control design for long duration autonomy," *Annual Reviews in Control*, vol. 46, pp. 1–7, 2018.

[129] M. Egerstedt and X. Hu, "A hybrid control approach to action coordination for mobile robots," *Automatica*, vol. 38, no. 1, pp. 125–130, 2002.

[130] J. Elith and J. R. Leathwick, "Species distribution models: Ecological explanation and prediction across space and time," *Annual Review of Ecology, Evolution, and Systematics*, vol. 40, pp. 677–697, 2009.

[131] Y. Emam, P. Glotfelter, and M. Egerstedt, "Robust barrier functions for a fully autonomous, remotely accessible swarm-robotics testbed," *IEEE Conference on Decision and Control*, pp. 3984–3990, 2019.

[132] H. B. Enderton, *A Mathematical Introduction to Logic*. Elsevier, San Diego, CA, 2001.

[133] J. Enright, M. Hilstad, A. Saenz-Otero, and D. W. Miller, "The SPHERES guest scientist program collaborative science on the ISS," *IEEE Aerospace Conference*, vol. 1, Mar. 2004.

[134] T. Eren, P. N. Belhumeur, B. D. O. Anderson, and A. S. Morse, "A framework for maintaining formations based on rigidity," *IFAC World Congress*, 2002.

[135] T. Eren, W. Whiteley, B. D. O. Anderson, A. S. Morse, and P. N. Belhumeur, "Information structures to secure control of rigid formations with leader-follower architecture," *American Control Conference*, pp. 2966–2971, 2005.

[136] R. Estes, *The Behavior Guide to African Mammals*, vol. 64. University of California Press, Berkeley, CA, 1991.

[137] R. F. Ewer, *Ethology of Mammals*. Springer, New York, NY, 2013.

[138] K. Fathian, D. I. Rachinskii, M. W. Spong, and N. R. Gans, "Globally asymptotically stable distributed control for distance and bearing based multi-agent formations," *American Control Conference*, pp. 4642–4648, Jul. 2016.

[139] K. Fathian, D. I. Rachinskii, T. H. Summers, and N. R. Gans, "Distributed control of cyclic formations with local relative position measurements," *IEEE Conference on Decision and Control*, 2016.

[140] J. A. Fax and R. M. Murray, "Information flow and cooperative control of vehicle formations," *IEEE Transactions on Automatic Control*, vol. 49, no. 9, pp. 1465–1476, 2004.

[141] A. F. Filippov, "Differential equations with discontinuous right-hand side," *Russian Academy of Sciences: Matematicheskii sbornik*, vol. 93, no. 1, pp. 99–128, 1960.

[142] J. Fink, M. A. Hsieh, and V. Kumar, "Multi-robot manipulation via caging in environments with obstacles," *IEEE International Conference on Robotics and Automation*, pp. 1471–1476, 2008.

[143] E. Fiorelli, N. E. Leonard, P. Bhatta, D. Paley, R. Bachmayer, and D. M. Fratantoni, "Multi-AUV control and adaptive sampling in Monterey Bay," *Journal of Oceanic Engineering*, vol. 31, no. 4, pp. 935–948, 2006.

[144] W. F. Fiske, "Superparasitism: An important factor in the natural control of insects," *Journal of Economic Entomology*, vol. 3, no. 1, pp. 88–97, 1910.

[145] A. D. Flouris, "Functional architecture of behavioural thermoregulation," *European Journal of Applied Physiology*, vol. 111, no. 1, pp. 1–8, 2011.

[146] F. Forget, F. Costard, and P. Lognonné, "*Spirit* and *Opportunity*: Wheels on Mars," in *Planet Mars: Story of Another World*. Springer Praxis Books, New York, NY, 2008.

[147] T. I. Fossen and A. M. Lekkas, "Direct and indirect adaptive integral line-of-sight path-following controllers for marine craft exposed to ocean currents," *International Journal of Adaptive Control and Signal Processing*, vol. 31, no. 4, pp. 445–463, 2017.

[148] V. Fourcassié, A. Dussutour, and J. L. Deneubourg, "Ant traffic rules," *Journal of Experimental Biology*, vol. 213, no. 14, pp. 2357–2363, 2010.

[149] D. Fox, W. Burgard, and S. Thrun, "The dynamic window approach to collision avoidance," *IEEE Robotics & Automation Magazine*, vol. 4, no. 1, pp. 23–33, 1997.

[150] M. Franceschelli, M. Egerstedt, and A. Giua, "Motion probes for fault detection and recovery in networked control systems," *American Control Conference*, pp. 4358–4363, 2008.

[151] T. B. Franklin, *Climates in Miniature: A Study of Micro-Climate and Environment*. Philosophical Library, New York, NY, 1955.

[152] R. Freeman and P. V. Kokotovic, *Robust Nonlinear Control Design: State-Space and Lyapunov Techniques*. Springer Science & Business Media, Berlin/Heidelberg, Germany, 2008.

[153] M. Freese, S. Singh, F. Ozaki, and N. Matsuhira, "Virtual robot experimentation platform V-REP: A versatile 3D robot simulator," *International Conference on Simulation, Modeling, and Programming for Autonomous Robots*, pp. 51–62, 2010.

[154] S. D. Fretwell and H. L. Lucas, "On territorial behavior and other factors influencing habitat distribution in birds. I. Theoretical development," *Acta Biotheoretica*, vol. 19, pp. 16–36, 1970.

[155] K. J. Gaston and J. I. Spicer, *Biodiversity: An Introduction*. John Wiley & Sons, Hoboken, NJ, 2013.

[156] B. Gates, "A robot in every home," *Scientific American*, vol. 296, no. 1, pp. 58–65, 2007.

[157] G. F. Gause, *The Struggle for Existence: A Classic of Mathematical Biology and Ecology*. Courier Dover Publications, Mineola, NY, 2019.

[158] Y. Girdhar and G. Dudek, "Modeling curiosity in a mobile robot for long-term autonomous exploration and monitoring," *Autonomous Robots*, vol. 40, no. 7, pp. 1267–1278, 2016.

[159] P. Glotfelter, I. Buckley, and M. Egerstedt, "Hybrid nonsmooth barrier functions with applications to provably safe and composable collision avoidance for robotic systems," *IEEE Robotics and Automation Letters*, vol. 4, no. 2, pp. 1303–1310, 2019.

[160] P. Glotfelter, J. Cortés, and M. Egerstedt, "Nonsmooth barrier functions with applications to multi-robot systems," *IEEE Control Systems Letters*, vol. 1, no. 2, pp. 310–315, 2017.

[161] P. Glotfelter, J. Cortés, and M. Egerstedt, "Boolean composability of constraints and control synthesis for multi-robot systems via nonsmooth control barrier functions," *IEEE Conference on Control Technology and Applications*, pp. 897–902, 2018.

[162] C. Godsil and G. F. Royle, *Algebraic Graph Theory*, vol. 207. Springer Science & Business Media, Berlin/Heidelberg, Germany, 2013.

[163] N. S. Goel, S. C. Maitra, and E. W. Montroll, "On the Volterra and other nonlinear models of interacting populations," *Reviews of Modern Physics*, vol. 43, no. 2, p. 231, 1971.

[164] R. Goodman and M. Herold, "Why maintaining tropical forests is essential and urgent for a stable climate," *Center for Global Development*, no. 385, 2014.

[165] D. M. Gordon, *Ant Encounters: Interaction Networks and Colony Behavior*. Princeton University Press, Princeton, NJ, 2010.

[166] R. Graham and J. Cortés, "Adaptive information collection by robotic sensor networks for spatial estimation," *IEEE Transactions on Automatic Control*, vol. 57, no. 6, pp. 1404–1419, 2012.

[167] G. Grisetti, R. Kummerle, C. Stachniss, and W. Burgard, "A tutorial on graph-based SLAM," *IEEE Intelligent Transportation Systems Magazine*, vol. 2, no. 4, pp. 31–43, 2010.

[168] S. Groeber, M. Vetter, B. Eckert, and H. J. Jodl, "Experimenting from a distance: A remotely controlled laboratory (RCL)," *European Journal of Physics*, vol. 28, no. 3, S127, 2007.

[169] M. J. Groom, G. K. Meffe, C. R. Carroll, and S. J. Andelman, *Principles of Conservation Biology*. Sinauer Associates, Sunderland, MA, 2006.

[170] D. Grunbaum, S. V. Viscido, and J. K. Parrish, "Extracting interactive control algorithms from group dynamics of schooling fish," in *Cooperative Control: Block Island Workshop on Cooperative Control*, Lecture Notes in Control and Information Science, vol. 309, edited by V. Kumar, N. Leonard, A. S. Morse, pp. 103–117. Springer Verlag, Berlin/Heidelberg, Germany, 2005.

[171] F. Halle and O. Pascal, "Biologie d'une canopee de foret equatorial II (Rapport de mission "radeau des cimes" Septembre-Decembre 1991, Campo, Cameroun)," *Institut Botanique*, 1992.

[172] W. D. Hamilton, "Geometry for the selfish herd," *Journal of Theoretical Biology*, vol. 31, no. 2, pp. 295–311, 1971.

[173] N. O. Handegard, K. M. Boswell, C. C. Ioannou, S. P. Leblanc, D. B. Tjostheim, and I. D. Couzin, "The dynamics of coordinated group hunting and collective information transfer among schooling prey," *Current Biology*, vol. 22, no. 13, pp. 1213–1217, 2012.

[174] F. Hapgood, "Chaotic robotics," *Wired*, vol. 2, 1994.

[175] M. Haque, M. Egerstedt, and C. F. Martin, "Sustainable group sizes for multi-agent search-andpatrol teams," *Mathematical Theory of Networks and Systems*, Budapest, Hungary, July 2010.

[176] M. Haque, M. Egerstedt, and A. Rahmani, "Multilevel coalition formation strategy for suppression of enemy air defenses missions," *Journal of Aerospace Information Systems*, vol. 10, no. 6, pp. 287–296, 2013.

[177] M. Haque, A. Rahmani, and M. Egerstedt, "Biologically inspired confinement of multi-robot systems," *International Journal of Bio-Inspired Computation*, vol. 3, no. 4, pp. 213–224, 2011.

[178] M. Haque, A. Rahmani, M. Egerstedt, and A. Yezzi, "Efficient foraging strategies in multi-agent systems through curve evolutions," *IEEE Transactions on Automatic Control*, vol. 59, no. 4, pp. 1036–1041, 2014.

[179] G. Hardin, "The competitive exclusion principle," *Science*, vol. 131, no. 3409, pp. 1292–1297, 1960.

[180] A. T. Hayes, "How many robots? Group size and efficiency in collective search tasks," *Distributed Autonomous Robotic Systems*, pp. 289–298, 2002.

[181] M. Héder, "From NASA to EU: The evolution of the TRL scale in public sector innovation," *The Innovation Journal*, vol. 22, no. 2, pp. 1–23, 2017.

[182] P. A. Henderson, *Practical Methods in Ecology*. John Wiley & Sons, Hoboken, NJ, 2009.

[183] J. Hespanha, "Uniform stability of switched linear systems: Extensions of LaSalle's invariance principle," *IEEE Transactions on Automatic Control*, vol. 49, no. 4, pp. 470–482, 2004.

[184] V. H. Heywood and R. T. Watson, *Global Biodiversity Assessment*. Cambridge University Press, Cambridge, UK, 1995.

[185] D. C. Hoaglin, F. Mosteller, and J. W. Tukey, *Understanding Robust and Exploratory Data Analysis*, vol. 3. Wiley, New York, NY, 1983.

[186] H. V. Hoell, J. T. Doyen, and A. H. Purcell, *Introduction to Insect Biology and Diversity*, 2nd ed. Oxford University Press, Oxford, UK, 1998.

[187] C. Hofner and G. Schmidt, "Path planning and guidance techniques for an autonomous mobile cleaning robot," *Robotics and Autonomous Systems*, vol. 14, no. 2–3, pp. 199–212, 1995.

[188] O. Holland, "Exploration and high adventure: The legacy of Grey Walter," *Philosophical Transactions of the Royal Society of London. Series A: Mathematical, Physical and Engineering Sciences*, vol. 361, no. 1811, pp. 2085–2121, 2003.

[189] G. A. Hollinger, S. Choudhary, P. Qarabaqi, C. Murphy, U. Mitra, G. S. Sukhatme, M. Stojanovic, H. Singh, and F. Hover, "Underwater data collection using robotic sensor networks," *IEEE Journal on Selected Areas in Communications*, vol. 30, no. 5, pp. 899–911, Jun. 2012.

[190] M. A. Hsieh, Á. Halász, S. Berman, and V. Kumar, "Biologically inspired redistribution of a swarm of robots among multiple sites," *Swarm Intelligence*, vol. 2, no. 2–4, pp. 121–141, 2008.

[191] M. A. Hubbe, "Life in the forest canopy," *BioResources*, vol. 8, no. 2, pp. 1508–1509, 2013.

[192] G. E. Hutchinson, "Quantitative biology," *Cold Spring Harbor Symposium*, vol. 22, pp. 415–427, 1957.

[193] Y. Igarashi, T. Hatanaka, M. Fujita, and M. W. Spong, "Passivity-based attitude synchronization in SE(3)," *IEEE Transactions on Control Systems Technology*, vol. 17, no. 5, pp. 1119–1134, 2009.

[194] A. Jadbabaie, J. Lin, and A. S. Morse, "Coordination of groups of mobile autonomous agents using nearest neighbor rules," *IEEE Transactions on Automatic Control*, vol. 48, no. 6, pp. 988–1001, 2003.

[195] K. H. Janstrup, *Road Safety Annual Report*. OECD Publishing, Paris, France, 2017.

[196] Jet Propulsion Laboratory, *Catalogued under Photo ID: PIA04413*, Public Domain image.

[197] M. Ji and M. Egerstedt, "Distributed coordination control of multi-agent systems while preserving connectedness," *IEEE Transactions on Robotics*, vol. 23, no. 4, pp. 693–703, 2007.

[198] A. Jimenez-Gonzalez, J. R. M. de Dios, and A. Ollero, "Testbeds for ubiquitous robotics: A survey," *Robotics and Autonomous Systems*, vol. 61, no. 12, pp. 1487–1501, 2013.

[199] D. Johnson, T. Stack, R. Fish, D. M. Flickinger, L. Stoller, R. Ricci, and J. Lepreau, "Mobile Emulab: A robotic wireless and sensor network testbed," *IEEE International Conference on Computer Communications*, pp. 1–12, 2006.

[200] H. G. Jones, *Plants and Microclimate: A Quantitative Approach to Environmental Plant Physiology*. Cambridge University Press, Cambridge, UK, 2013.

[201] J. L. Jones, "Robots at the tipping point: The road to iRobot Roomba," *IEEE Robotics and Automation Magazine*, vol. 13, no. 1, pp. 76–78, Mar. 2006.

[202] L. P. Kaelbling, M. L. Littman, and A. W. Moore, "Reinforcement learning: A survey," *Journal of Artificial Intelligence Research*, vol. 4, pp. 237–285, 1996.

[203] N. Kamra and N. Ayanian, "A mixed integer programming model for timed deliveries in multirobot systems," *IEEE International Conference on Automation Science and Engineering*, pp. 612–617, 2015.

[204] A. Kaplan, N. Kingry, P. Uhing, and R. Dai, "Time-optimal path planning with power schedules for a solar-powered ground robot," *IEEE Transactions on Automation Science and Engineering*, vol. 14, no. 2, pp. 1235–1244, 2017.

[205] N. Kashiri, A. Abate, S. J. Abram, A. Albu-Schaffer, P. J. Clary, M. Daley, S. Faraji, R. Furnemont, M. Garabini, H. Geyer, and A. M. Grabowski, "An overview on principles for energy efficient robot locomotion," *Frontiers in Robotics and AI*, vol. 5, p. 129, 2018.

[206] M. Kearney and W. Porter, "Mechanistic niche modelling: Combining physiological and spatial data to predict species' ranges," *Ecology Letters*, vol. 12, no. 4, pp. 334–350, 2009.

[207] P. A. Keddy, *Plants and Vegetation: Origins, Processes, Consequences*. Cambridge University Press, Cambridge, UK, 2007.

[208] L. H. Keel and S. P. Bhattacharyya, "Robust, fragile, or optimal?," *IEEE Transactions on Automatic Control*, vol. 42, no. 8, pp. 1098–1105, Aug. 1997.

[209] H. K. Khalil, *Nonlinear Systems*, 3rd ed. Pearson, New York, NY, 2001.

[210] S. M. Khansari-Zadeh and A. Billard, "Learning control Lyapunov function to ensure stability of dynamical system-based robot reaching motions," *Robotics and Autonomous Systems*, vol. 62, no. 6, pp. 752–765, 2014.

[211] E. Klavins, R. Ghrist, and D. Lipsky, "Graph grammars for self assembling robotic systems," *IEEE International Conference on Robotics and Automation*, vol. 5, pp. 5293–5300, 2004.

[212] E. Klavins and D. E. Koditschek, "A formalism for the composition of concurrent robot behaviors," *IEEE International Conference on Robotics and Automation*, vol. 4, pp. 3395–3402, 2000.

[213] M. Kloetzer and C. Belta, "Temporal logic planning and control of robotic swarms by hierarchical abstractions," *IEEE Transactions on Robotics*, vol. 23, no. 2, pp. 320–330, 2007.

[214] N. Koenig and A. Howard, "Design and use paradigms for Gazebo, an open-source multi-robot simulator," *IEEE/RSJ International Conference on Intelligent Robots and Systems*, vol. 3, pp. 2149–2154, Sep. 2004.

[215] L. P. Koh, R. R. Dunn, N. S. Sodhi, R. K. Colwell, H. C. Proctor, and V. S. Smith, "Species coextinctions and the biodiversity crisis," *Science*, vol. 305, no. 5690, pp. 1632–1634, 2004.

[216] J. R. Kok, M. T. J. Spaan, and N. Vlassis, "Multi-robot decision making using coordination graphs," *International Conference on Advanced Robotics*, vol. 3, pp. 1124–1129, 2003.

[217] V. R. Kompella, M. Stollenga, M. Luciw, and J. Schmidhuber, "Continual curiosity-driven skill acquisition from high-dimensional video inputs for humanoid robots," *Artificial Intelligence*, vol. 247, pp. 313–335, Jun. 2017.

[218] A. Krause, A. Singh, and C. Guestrin, "Near-optimal sensor placements in Gaussian processes: Theory, efficient algorithms and empirical studies," *Journal of Machine Learning Research*, vol. 9, pp. 235–284, Feb. 2008.

[219] B. Kuipers and Y. T. Byun, "A robot exploration and mapping strategy based on a semantic hierarchy of spatial representations," *Robotics and Autonomous Systems*, vol. 8, no. 1, pp. 47–63, 1991.

[220] A. Kwok and S. Martinez, "Deployment algorithms for a power-constrained mobile sensor network," *International Journal of Robust and Nonlinear Control*, vol. 20, no. 7, pp. 745–763, 2010.

[221] A. Kwok and S. Martinez, "Coverage maximization with autonomous agents in fast flow environments," *Journal of Optimization Theory and Applications*, vol. 155, no. 2, pp. 986–1007, 2012.

[222] T. H. Labella, M. Dorigo, and J. L. Deneubourg, "Self-organised task allocation in a group of robots," *Distributed Autonomous Robotic Systems*, pp. 389–398, 2007.

[223] J. Lancaster and B. J. Downes, *Aquatic Entomology*. Oxford University Press, Oxford, UK, 2013.

[224] C. G. Langton, *Artificial Life: An Overview*. MIT Press, Cambridge, MA, 1997.

[225] S. M. LaValle, *Planning Algorithms*. Cambridge University Press, Cambridge, UK, 2006.

[226] K. Laventall and J. Cortés, "Coverage control by multi-robot networks with limited-range anisotropic sensors," *International Journal of Control*, vol. 82, no. 6, pp. 1113–1121, 2009.

[227] J. Lawton, R. Beard, and B. Young, "A decentralized approach to formation maneuvers," *IEEE Transactions on Robotics and Automation*, vol. 19, no. 6, pp. 933–941, 2003.

[228] H. J. LeBlanc, H. Zhang, X. Koutsoukos, and S. Sundaram, "Resilient asymptotic consensus in robust networks," *IEEE Journal on Selected Areas in Communications*, vol. 31, no. 4, pp. 766–781, 2013.

[229] S. Lee, Y. Diaz-Mercado, and M. Egerstedt, "Multirobot control using time-varying density functions," *IEEE Transactions on Robotics*, vol. 31, no. 2, pp. 489–493, 2015.

[230] G. Leitmann, *The Calculus of Variations and Optimal Control: An Introduction*. Springer Science & Business Media, Berlin/Heidelberg, Germany, 2013.

[231] N. Lemmens, S. De Jong, K. Tuyls, and A. Nowe, "Bee behaviour in multi-agent systems (a bee foraging algorithm)," in *Adaptive Agents and Multi-agent Systems III*, Lecture Notes in Computer Science, edited by K. Tuyls, A. Nowe, Z. Guessoum, and D. Kudenko, pp. 145–156. Springer Verlag, Berlin/Heidelberg, Germany, 2005.

[232] N. E. Leonard and E. Fiorelli, "Virtual leaders, artificial potentials and coordinated control of groups," *IEEE Conference on Decision and Control*, 2001.

[233] N. E. Leonard, D. A. Paley, R. E. Davis, D. M. Fratantoni, F. Lekien, and F. Zhang, "Coordinated control of an underwater glider fleet in an

adaptive ocean sampling field experiment in Monterey Bay," *Journal of Field Robotics*, vol. 27, no. 6, pp. 718–740, 2010.

[234] K. Lerman and A. Galstyan, "Mathematical model of foraging in a group of robots: Effect of interference," *Autonomous Robots*, vol. 13, no. 2, pp. 127–141, 2002.

[235] A. Li, L. Wang, P. Pierpaoli, and M. Egerstedt, "Formally correct composition of coordinated behaviors using control barrier certificates," *IEEE/RSJ International Conference on Intelligent Robots and Systems*, pp. 3723–3729, 2018.

[236] W. Li, S. Chow, M. Egerstedt, J. Lu, and H. Zhou, "Method of evolving junctions: A new approach to optimal path-planning in 2D environments with moving obstacles," *International Journal of Robotics Research*, vol. 36, no. 4, pp. 403–413, May 2017.

[237] Z. Li and J. F. Canny, *Nonholonomic Motion Planning*. Springer Science & Business Media, Berlin/Heidelberg, Germany, 2012.

[238] D. Liberzon, *Calculus of Variations and Optimal Control Theory: A Concise Introduction*. Princeton University Press, Princeton, NJ, 2011.

[239] J. Lin, A. S. Morse, and B. D. O. Anderson, "The multi-agent rendezvous problem. Part 1: The synchronous case," *SIAM Journal on Control and Optimization*, vol. 46, no. 6, pp. 2096–2119, 2007.

[240] P. Lin, K. Abney, and G. A. Bekey, *Robot Ethics: The Ethical and Social Implications of Robotics*, Intelligent Robotics and Autonomous Agents series. MIT Press, Cambridge, MA, 2012.

[241] P. Lin, K. Abney, and R. Jenkins, *Robot Ethics 2.0: From Autonomous Cars to Artificial Intelligence*. Oxford University Press, Oxford, UK, 2017.

[242] L. Liu and N. Michael, "Energy-aware aerial vehicle deployment via bipartite graph matching," *IEEE International Conference on Unmanned Aircraft Systems*, pp. 189–194, 2014.

[243] S. Y. Liu and K. Hedrick, "The application of domain of danger in autonomous agent team and its effect on exploration efficiency," *American Control Conference*, pp. 4111–4116, 2011.

[244] W. Liu, Z. Wang, X. Liu, N. Zeng, Y. Liu, and F. E. Alsaadi, "A survey of deep neural network architectures and their applications," *Neurocomputing*, vol. 234, pp. 11–26, 2017.

[245] S. Lloyd, "Least squares quantization in PCM," *IEEE Transactions on Information Theory*, vol. 28, no. 2, pp. 129–137, 1982.

[246] K. J. Locey and J. T. Lennon, "Scaling laws predict global microbial diversity," *Proceedings of the National Academy of Sciences*, vol. 113, no. 21, pp. 5970–5975, 2016.

[247] J. T. Longino and R. K. Colwell, "Biodiversity assessment using structured inventory: Capturing the ant fauna of a tropical rain forest," *Ecological Applications*, vol. 7, no. 4, pp. 1263–1277, 1997.

[248] K. Lorenz, *The Foundations of Ethology*. Springer Science & Business Media, Berlin/Heidelberg, Germany, 2013.

[249] M. D. Lowman and B. Bouricius, "The construction of platforms and bridges for forest canopy access," *Selbyana*, vol. 16, no. 2, pp. 179–184, 1995.

[250] M. D. Lowman and T. D. Schowalter, "Plant science in forest canopies: The first 30 years of advances and challenges (1980–2010)," *New Phytologist*, vol. 194, no. 1, pp. 12–27, 2012.

[251] D. G. Luenberger, *Optimization by Vector Space Methods*. John Wiley & Sons, Hoboken, NJ, 1997.

[252] D. G. Luenberger and Y. Ye, *Linear and Nonlinear Programming*. Springer, New York, NY, 1984.

[253] N. Lynch, *Distributed Algorithms*. Elsevier, Burlington, MA, 1996.

[254] R. H. MacArthur and E. R. Pianka, "On optimal use of a patchy environment," *The American Naturalist*, vol. 100, no. 916, pp. 603–609, Dec. 1966.

[255] D. I. MacKenzie and W. L. Kendall, "How should detection probability be incorporated into estimates of relative abundance?," *Ecology*, vol. 83, no. 9, pp. 2387–2393, 2002.

[256] J. D. Madden, R. C. Arkin, and D. R. MacNulty, "Multi-robot system based on model of wolf hunting behavior to emulate wolf and elk interactions," *IEEE International Conference on Robotics and Biomimetics*, pp. 1043–1050, 2010.

[257] M. W. Maimone, "Global path planning on board the Mars exploration rovers," May 2008.

[258] M. W. Maimone, P. C. Leger, and J. J. Biesiadecki, "Overview of the Mars exploration rovers' autonomous mobility and vision capabilities," *IEEE International Conference on Robotics and Automation*, May 2007.

[259] J. Mann, R. C. Connor, P. L. Tyack, and H. Whitehead, *Cetacean Societies: Field Studies of Dolphins and Whales*. University of Chicago Press, Chicago, IL, 2000.

[260] A. Marino, L. Parker, G. Antonelli, and F. Caccavale, "Behavioral control for multi-robot perimeter patrol: A finite state automata approach," *IEEE International Conference on Robotics and Automation*, pp. 831–836, 2009.

[261] J. A. Marshall, M. E. Broucke, and B. A. Francis, "Formations of vehicles in cyclic pursuit," *IEEE Transactions on Automatic Control*, vol. 49, no. 11, pp. 1963–1974, 2004.

[262] P. Martin and M. Egerstedt, "Hybrid systems tools for compiling controllers for cyber-physical systems," *Discrete Event Dynamic Systems*, vol. 22, no. 1, pp. 101–119, 2012.

[263] S. Martinez, J. Cortés, and F. Bullo, "Motion coordination with distributed information," *IEEE Control Systems Magazine*, vol. 27, no. 4, pp. 75–88, 2007.

[264] A. Martinoli, A. J. Ijspeert, and L. M. Gambardella, "A probabilistic model for understanding and comparing collective aggregation mechanisms," *European Conference on Artificial Life*, pp. 575–584, 1999.

[265] G. Mathew and I. Mezić, "Metrics for ergodicity and design of ergodic dynamics for multi-agent systems," *Physica D: Nonlinear Phenomena*, vol. 240, no. 4, pp. 432–442, 2011.

[266] N. Mathew, S. L. Smith, and S. L. Waslander, "Multirobot rendezvous planning for recharging in persistent tasks," *IEEE Transactions on Robotics*, vol. 31, no. 1, pp. 128–142, 2015.

[267] S. Mayya, P. Pierpaoli, and M. Egerstedt, "Voluntary retreat for decentralized interference reduction in robot swarms," *IEEE International Conference on Robotics and Automation*, pp. 9667–9673, 2019.

[268] S. Mayya, P. Pierpaoli, G. Nair, and M. Egerstedt, "Localization in densely packed swarms using interrobot collisions as a sensing

modality," *IEEE Transactions on Robotics*, vol. 35, no. 1, pp. 21–34, 2018.

[269] S. Mayya, S. Wilson, and M. Egerstedt, "Closed-loop task allocation in robot swarms using inter-robot encounters," *Swarm Intelligence*, vol. 13, no. 2, pp. 115–143, 2019.

[270] D. J. McFarland, *Motivational Control Systems Analysis*. Academic Press, Cambridge, MA, 1974.

[271] J. M. McNew, E. Klavins, and M. Egerstedt, "Solving coverage problems with embedded graph grammars," in *Hybrid Systems: Computation and Control*, edited by A. Bemporad, A. Bicchi, and G. Buttazzo, pp. 413–427. Springer, Berlin, 2007.

[272] P. N. Mehra and K. S. Bawa, "B-Chromosomes in some Himalayan hardwoods," *Chromosoma*, vol. 25, no. 1, pp. 90–95, 1968.

[273] D. Mellinger, A. Kushleyev, and V. Kumar, "Mixed-integer quadratic program trajectory generation for heterogeneous quadrotor teams," *IEEE International Conference on Robotics and Automation*, pp. 477–483, 2012.

[274] Merriam-Webster, *Collegiate Dictionary*. Springfield, MA, 2004.

[275] M. Mesbahi, "State-dependent graphs," *IEEE Conference on Decision and Control*, vol. 3, pp. 3058–3063, 2003.

[276] M. Mesbahi and M. Egerstedt, *Graph Theoretic Methods in Multiagent Networks*. Princeton University Press, Princeton, NJ, 2010.

[277] N. Michael and V. Kumar, "Controlling shapes of ensembles of robots of finite size with nonholonomic constraints," *Robotics: Science and Systems*, Jun. 2008.

[278] O. Michel, "Webots: Professional mobile robot simulation," *Journal of Advanced Robotics Systems*, vol. 1, no. 1, pp. 39–42, 2004.

[279] L. M. Miller and T. D. Murphey, "Trajectory optimization for continuous ergodic exploration," *American Control Conference*, pp. 4196–4201, 2013.

[280] J. Mirkovic and T. Benzel, "Teaching cybersecurity with DeterLab," *IEEE Security and Privacy Magazine*, vol. 10, no. 1, pp. 73–76, Jan. 2012.

[281] D. Mitchell, M. Corah, N. Chakraborty, K. Sycara, and N. Michael, "Multi-robot long-term persistent coverage with fuel constrained

robots," *IEEE International Conference on Robotics and Automation*, pp. 1093–1099, 2015.

[282] S. Mohan, A. Saenz-Otero, S. Nolet, D. W. Miller, and S. Sell, "SPHERES flight operations testing and execution," *Acta Astronautica*, vol. 65, no. 7–8, pp. 1121–1132, 2009.

[283] C. Mora, D. Tittensor, S. Adl, A. Simpson, and B. Worm, "How many species are there on Earth and in the ocean?," *PLoS Biology*, vol. 9, no. 8, e1001127, 2011.

[284] H. Moravec, "Locomotion, vision and intelligence," in *Robotics Research*, edited by M. Brady and R. Paul, pp. 215–224. MIT Press, Cambridge, MA, 1984.

[285] L. Moreau, "Stability of multiagent systems with time-dependent communication links," *IEEE Transactions on Automatic Control*, vol. 50, no. 2, pp. 169–182, 2005.

[286] N. Morozovsky and T. Bewley, "SkySweeper: A low DOF, dynamic high wire robot," *IEEE/RSJ International Conference on Intelligent Robots and Systems*, pp. 2339–2344, 2013.

[287] M. Morris and S. Tosunoglu, "Survey of rechargeable batteries for robotic applications," *Florida Conference on Recent Advances in Robotics*, May 2012.

[288] K. P. Murphy, *Machine Learning: A Probabilistic Perspective*. MIT Press, Cambridge, MA, 2012.

[289] S. Naeem and S. Li, "Biodiversity enhances ecosystem reliability," *Nature*, vol. 390, no. 6659, pp. 507–509, 1997.

[290] M. Nagumo, "Über die lage der integralkurven gewöhnlicher differentialgleichungen," in *Proceedings of the Physico-Mathematical Society of Japan*, vol. 24, pp. 551–559, 1942.

[291] N. Napp and E. Klavins, "A compositional framework for programming stochastically interacting robots," *International Journal of Robotics Research*, vol. 30, no. 6, pp. 713–729, 2011.

[292] NASA, "Mars exploration rovers," https://mars.nasa.gov/mer/.

[293] NASA, "Technology readiness level definitions," https://www.nasa.gov /pdf/458490main_TRL_Definitions.pdf.

[294] M. Nayyerloo, X. Q. Chen, W. Wang, and J. G. Chase, "Cable-climbing robots for power transmission lines inspection," in *Mobile Robots-State of the Art in Land, Sea, Air, and Collaborative Missions*, edited by X. Chen and Y. Q. Chen. InTech, London, UK, 2009.

[295] M. Nayyerloo, S. Yeganehparast, A. Barati, and M. Foumani, "Mechanical implementation and simulation of Monolab, a mobile robot for inspection of power transmission lines," *International Journal of Advanced Robotic Systems*, vol. 4, no. 3, pp. 381–386, 2007.

[296] R. W. Nelson, S. G. Lever, D. M. Gaines, J. A. Herman, P. P. Hwang, S. L. Laubach, M. M. Pack, and S. F. Peters, "On low power operations during *Spirit*'s third winter on Mars," *IEEE Aerospace Conference*, Mar. 2009.

[297] Q. Nguyen, A. Hereid, J. W. Grizzle, A. D. Ames, and K. Sreenath, "3D dynamic walking on stepping stones with control barrier functions," *IEEE Conference on Decision and Control*, pp. 827–834, 2016.

[298] Q. Nguyen and K. Sreenath, "Exponential control barrier functions for enforcing high relative-degree safety-critical constraints," *American Control Conference (ACC)*, pp. 322–328, 2016.

[299] K. Nishimura, "Foraging in an uncertain environment: Patch exploitation," *Journal of Theoretical Biology*, vol. 156, no. 1, pp. 91–111, 1992.

[300] G. Notomista, X. Cai, J. Yamauchi, and M. Egerstedt, "Passivity-based decentralized control of multi-robot systems with delays using control barrier functions," *International Symposium on Multi-Robot and Multi-Agent Systems*, pp. 231–237, 2019.

[301] G. Notomista and M. Egerstedt, "Coverage control for wire-traversing robots," *IEEE International Conference on Robotics and Automation*, pp. 1–6, 2018.

[302] G. Notomista and M. Egerstedt, "Persistification of robotic tasks," *IEEE Transactions on Control Systems Technology*, vol. 29, no. 2, pp. 756–767, Mar. 2021.

[303] G. Notomista, Y. Emam, and M. Egerstedt, "The SlothBot: A novel design for a wire-traversing robot," *Robotics and Automation Letters*, vol. 4, no. 2, pp. 1993–1998, Feb. 2019.

[304] G. Notomista, S. Ruf, and M. Egerstedt, "Persistification of robotic tasks using control barrier functions," *Robotics and Automation Letters*, vol. 3, no. 2, pp. 758–763, Apr. 2018.

[305] G. Notomista, M. Santos, S. Hutchinson, and M. Egerstedt, "Sensor coverage control using robots constrained to a curve," *IEEE International Conference on Robotics and Automation*, pp. 3252–3258, 2019.

[306] G. Notomista and M. Egerstedt, "Constraint-driven coordinated control of multi-robot systems," *American Control Conference*, pp. 1990–1996, 2019.

[307] M. J. O'Brien and R. C. Arkin, "An artificial circadian system for a slow and persistent robot," *International Conference on Simulation of Adaptive Behavior*, pp. 149–161, 2018.

[308] E. P. Odum and G. W. Barrett, *Fundamentals of Ecology*. Cengage Learning, Boston, MA, 2004.

[309] R. O'Flaherty and M. Egerstedt, "Optimal exploration in unknown environments," *IEEE/RSJ International Conference on Intelligent Robots and Systems*, pp. 5796–5801, 2015.

[310] P. Ögren, A. Backlund, T. Harryson, L. Kristensson, and P. Stensson, "Autonomous UCAV strike missions using behavior control Lyapunov functions," *AIAA Guidance, Navigation, and Control Conference and Exhibit*, p. 6197, 2006.

[311] P. Ögren, M. Egerstedt, and X. Hu, "A control Lyapunov function approach to multi-agent coordination," *IEEE Transactions on Robotics and Automation*, vol. 18, no. 5, pp. 847–851, 2002.

[312] P. Ögren and N. E. Leonard, "A tractable convergent dynamic window approach to obstacle avoidance," *IEEE/RSJ International Conference on Intelligent Robots and Systems*, vol. 1, pp. 595–600, 2002.

[313] P. Ögren and N. E. Leonard, "A convergent dynamic window approach to obstacle avoidance," *IEEE Transactions on Robotics*, vol. 21, no. 2, pp. 188–195, 2005.

[314] A. Okabe and A. Suzuki, "Locational optimization problems solved through Voronoi diagrams," *European Journal of Operational Research*, vol. 98, no. 3, pp. 445–456, 1997.

[315] R. Olfati-Saber, "Near-identity diffeomorphisms and exponential tracking and stabilization of first-order nonholonomic SE(2) vehicles," *American Control Conference*, vol. 6, pp. 4690–4695, 2002.

[316] R. Olfati-Saber, J. A. Fax, and R. M. Murray, "Consensus and cooperation in networked multi-agent systems," *Proceedings of the IEEE*, vol. 95, no. 1, pp. 215–233, 2007.

[317] R. Olfati-Saber and R. M. Murray, "Consensus problems in networks of agents with switching topology and time-delays," *IEEE Transactions on Automatic Control*, vol. 49, no. 9, pp. 1520–1533, 2004.

[318] R. Olfati-Saber, "Flocking for multi-agent dynamic systems: Algorithms and theory," *IEEE Transactions on Automatic Control*, vol. 51, no. 3, pp. 401–420, 2006.

[319] A. Ollero, S. Lacroix, L. Merino, J. Gancet, J. Wiklund, V. Remus, I. V. Perez, L. G. Gutierrez, D. X. Viegas, and M. A. G. Benitez, "Multiple eyes in the skies: Architecture and perception issues in the COMETS unmanned air vehicles project," *Robotics & Automation Magazine*, vol. 12, no. 2, pp. 46–57, 2005.

[320] E. Omerdic, D. Toal, and G. Dooly, "Remote presence: Powerful tool for promotion, education and research in marine robotics," *IEEE St. John's OCEANS*, 2015.

[321] E. Ostergaard, G. Sukhatme, and M. Mataric, "Emergent bucket brigading," *Autonomous Agents*, vol. 37, pp. 2219–2223, 2001.

[322] B. E. Paden and S. S. Sastry, "A calculus for computing Filippov's differential inclusion with application to the variable structure control of robot manipulators," *IEEE Conference on Decision and Control*, pp. 578–582, Dec. 1986.

[323] V. I. Pajunen, "The influence of population density on the territorial behaviour of Leucrrhinia rubicunda," *Annales zoologici fennici*, vol. 3, no. 1, pp. 40–52, 1966.

[324] C. S. Park, M. J. Tahk, and H. Bang, "Multiple aerial vehicle formation using swarm intelligence," *AIAA Guidance, Navigation, and Control Conference and Exhibit*, p. 5729, 2003.

[325] H. Park and S. Hutchinson, "An efficient algorithm for fault-tolerant rendezvous of multi-robot systems with controllable sensing

range," *IEEE International Conference on Robotics and Automation*, pp. 358–365, May 2016.

[326] H. Park and S. Hutchinson, "Fault-tolerant rendezvous of multirobot systems," *IEEE Transactions on Robotics*, vol. 33, no. 3, pp. 565–582, 2017.

[327] G. G. Parker, "Structure and microclimate of forest canopies," in *Forest Canopies*, edited by M. D. Lowman and N. M. Nadkarni. Academic Press, Cambridge, MA, 1995.

[328] J. N. Pauli, J. E. Mendoza, S. A. Steffan, C. C. Carey, P. J. Weimer, and M. Z. Peery, "A syndrome of mutualism reinforces the lifestyle of a sloth," *Proceedings of the Royal Society B: Biological Sciences*, vol. 281, no. 1778, p. 20133006, 2014.

[329] L. Paull, J. Tani, H. Ahn, J. Alonso-Mora, L. Carlone, M. Cap, Y. F. Chen, C. Choi, J. Dusek, Y. Fang, D. Hoehener, S. Liu, M. Novitzky, I. F. Okuyama, J. Pazis, G. Rosman, V. Varricchio, H. Wang, D. Yershov, H. Zhao, M. Benjamin, C. Carr, M. Zuber, S. Karaman, E. Frazzoli, D. Del Vecchio, D. Rus, J. How, J. Leonard, and A. Censi, "Duckietown: An open, inexpensive and flexible platform for autonomy education and research," *IEEE International Conference on Robotics and Automation*, pp. 1497–1504, May 2017.

[330] T. P. Pavlic and K. M. Passino, "Generalizing foraging theory for analysis and design," *The International Journal of Robotics Research*, vol. 30, no. 5, pp. 505–523, 2011.

[331] R. Pearl and L. J. Reed, "On the rate of growth of the population of the United States since 1790 and its mathematical representation," *Proceedings of the National Academy of Sciences of the United States of America*, vol. 6, no. 6, p. 275, 1920.

[332] M. Z. Peery and J. N. Pauli, "Shade-grown cacao supports a self-sustaining population of two-toed but not three-toed sloths," *Journal of Applied Ecology*, vol. 51, no. 1, pp. 162–170, 2014.

[333] K. E. Petersen, *Ergodic Theory*. Cambridge University Press, Cambridge, UK, 1989.

[334] L. Petersson, D. Austin, and H. I. Christensen, "DCA: A distributed control architecture for robotics," *IEEE/RSJ International Conference on Intelligent Robots and Systems*, vol. 4, pp. 2361–2368, 2001.

[335] P. Petrovic and R. Balogh, "Deployment of remotely-accessible robotics laboratory," *International Journal of Online Engineering*, vol. 8, no. S2, pp. 31–35, 2012.

[336] A. J. Phillips, J. M. Major, and G. R. Bartlett, *Line inspection robot and system*, US Patent 8,666,553, 2014.

[337] D. Pickem, P. Glotfelter, L. Wang, M. Mote, A. D. Ames, E. Feron, and M. Egerstedt, "The Robotarium: A remotely accessible swarm robotics research testbed," *IEEE International Conference on Robotics and Automation*, pp. 1699–1706, 2017.

[338] D. Pickem, M. Lee, and M. Egerstedt, "The GRITSBot in its natural habitat—A multi-robot testbed," *IEEE International Conference on Robotics and Automation*, pp. 4062–4067, 2015.

[339] A. Pocheville, "The ecological niche: History and recent controversies," in *Handbook of Evolutionary Thinking in the Sciences*, edited by T. Heams, P. Huneman, G. Lecointre, and M. Silberstein, pp. 547–586. Springer, Berlin/Heidelberg, Germany, 2015.

[340] N. Pouliot and S. Montambault, "Geometric design of the LineScout, a teleoperated robot for power line inspection and maintenance," *IEEE International Conference on Robotics and Automation*, pp. 3970–3977, 2008.

[341] R. B. Primack, *Essentials of Conservation Biology*. Sinauer Associates, Sunderland, MA, 1993.

[342] A. Prorok, M. A. Hsieh, and V. Kumar, "The impact of diversity on optimal control policies for heterogeneous robot swarms," *IEEE Transactions on Robotics*, vol. 33, no. 2, pp. 346–358, 2017.

[343] K. Pryor and K. S. Norris, *Dolphin Societies: Discoveries and Puzzles*. University of California Press, Berkeley, CA, 1991.

[344] D. Purves, G. J. Augustine, D. Fitzpatrick, W. C. Hall, A. S. LaMantia, J. O. McNamara, and S. M. Williams, *Neuroscience*. Sinauer Associates, Sunderland, MA, 2004.

[345] N. Quijano and K. M. Passino, "The ideal free distribution: Theory and engineering application," *IEEE Transactions on Systems, Man, and Cybernetics, Part B (Cybernetics)*, vol. 37, no. 1, pp. 154–165, 2007.

[346] I. Rahwan, M. Cebrian, N. Obradovich, J. Bongard, J. F. Bonnefon, C. Breazeal, J. W. Crandall, N. A. Christakis, I. D. Couzin, and

M. O. Jackson, "Machine behaviour," *Nature*, vol. 568, no. 7753, pp. 477–486, 2019.

[347] A. Ramezani, S. J. Chung, and S. Hutchinson, "A biomimetic robotic platform to study flight specializations of bats," *Science Robotics*, vol. 2, no. 3, 2017.

[348] N. Ranasinghe, J. Everist, and W. M. Shen, "Modular robot climbers," *IEEE/RSJ International Conference on Intelligent Robots and Systems*, 2007.

[349] C. R. Rao, "Diversity and dissimilarity coefficients: A unified approach," *Theoretical Population Biology*, vol. 21, no. 1, pp. 24–43, 1982.

[350] C. E. Rasmussen, "Gaussian processes in machine learning," in *Advanced Lectures on Machine Learning*, Lecture Notes in Computer Science, edited by O. Bousquet, U. von Luxburg, G. Rätsch, pp. 63–71. Springer Verlag, Berlin/Heidelberg, Germany, 2003.

[351] H. Ravanbakhsh and S. Sankaranarayanan, "Learning Lyapunov (potential) functions from counterexamples and demonstrations," *Robotics: Science and Systems*, Cambridge, MA, July 2017.

[352] D. Raychaudhuri, I. Seskar, M. Ott, S. Ganu, K. Ramachandran, H. Kremo, R. Siracusa, H. Liu, and M. Singh, "Overview of the ORBIT radio grid testbed for evaluation of next-generation wireless network protocols," *IEEE Wireless Communications and Networking Conference*, vol. 3, pp. 1664–1669, Mar. 2005.

[353] J. B. Reece, L. A. Urry, M. L. Cain, S. A. Wasserman, P. V. Minorsky, and R. B. Jackson, *Campbell Biology*. Pearson, Boston, MA, 2014.

[354] J. Reher, E. A. Cousineau, A. Hereid, C. M. Hubicki, and A. D. Ames, "Realizing dynamic and efficient bipedal locomotion on the humanoid robot DURUS," *IEEE International Conference on Robotics and Automation*, pp. 1794–1801, 2016.

[355] W. Ren and R. W. Beard, *Distributed Consensus in Multi-Vehicle Cooperative Control*. Springer Verlag, London, UK, 2008.

[356] C. W. Reynolds, "Flocks, herds and schools: A distributed behavioral model," *Conference on Computer Graphics and Interactive Techniques*, pp. 25–34, 1987.

[357] R. E. Ricklefs, *The Economy of Nature*. Macmillan, New York, NY, 2008.

[358] R. T. Rockafellar, *Convex Analysis*. Princeton University Press, Princeton, NJ, 1970.

[359] M. O. Rödel and R. Ernst, "Measuring and monitoring amphibian diversity in tropical forests. I. An evaluation of methods with recommendations for standardization," *Ecotropica*, vol. 10, no. 1, pp. 1–14, 2004.

[360] N. J. Rosenberg, B. L. Blad, and S. B. Verma, *Microclimate: The Biological Environment*. John Wiley & Sons, Hoboken, NJ, 1983.

[361] J. K. Rosenblatt, "DAMN: A distributed architecture for mobile navigation," *Journal of Experimental & Theoretical Artificial Intelligence*, vol. 9, no. 2–3, pp. 339–360, 1997.

[362] S. I. Roumeliotis and M. J. Mataric, "Small-world networks of mobile robots," *AAAI/IAAI*, p. 1093, 2000.

[363] I. Saha, R. Ramaithitima, V. Kumar, G. J. Pappas, and S. A. Seshia, "Automated composition of motion primitives for multi-robot systems from safe LTL specifications," *IEEE/RSJ International Conference on Intelligent Robots and Systems*, pp. 1525–1532, 2014.

[364] M. Santos and M. Egerstedt, "Coverage control for multi-robot teams with heterogeneous sensing capabilities using limited communications," *IEEE/RSJ International Conference on Intelligent Robots and Systems*, pp. 5313–5319, 2018.

[365] M. Santos, S. Mayya, G. Notomista, and M. Egerstedt, "Decentralized minimum-energy coverage control for time-varying density functions," *IEEE International Symposium on Multi-Robot and Multi-Agent Systems*, 2019.

[366] S. S. Sastry, *Nonlinear Systems: Analysis, Stability, and Control*. Springer Science & Business Media, Berlin/Heidelberg, Germany, 2013.

[367] J. Sattar, G. Dudek, O. Chiu, I. Rekleitis, P. Giguere, A. Mills, N. Plamondon, C. Prahacs, Y. Girdhar, M. Nahon, and J. P. Lobos, "Enabling autonomous capabilities in underwater robotics," *IEEE/RSJ International Conference on Intelligent Robots and Systems*, Sep. 2008.

[368] J. Sawada, K. Kusumoto, Y. Maikawa, T. Munakata, and Y. Ishikawa, "A mobile robot for inspection of power transmission lines," *IEEE Transactions on Power Delivery*, vol. 6, no. 1, pp. 309–315, 1991.

[369] S. Schaal, "Is imitation learning the route to humanoid robots?," *Trends in Cognitive Sciences*, vol. 3, no. 6, pp. 233–242, 1999.

[370] M. Schneider-Fontan and M. J. Mataric, "Territorial multi-robot task division," *IEEE Transactions on Robotics and Automation*, vol. 14, no. 5, pp. 815–822, 1998.

[371] M. D. Schuresko and J. Cortés, "Distributed tree rearrangements for reachability and robust connectivity," *SIAM Journal on Control and Optimization*, vol. 50, no. 5, pp. 2588–2620, 2012.

[372] R. J. Schusterman, J. A. Thomas, F. G. Wood, and R. Schusterman, *Dolphin Cognition and Behavior: A Comparative Approach*. Psychology Press, Hove, UK, 2013.

[373] M. L. Seto, *Marine Robot Autonomy*. Springer Science & Business Media, Berlin/Heidelberg, Germany, 2012.

[374] T. Setter and M. Egerstedt, "Energy-constrained coordination of multi-robot teams," *IEEE Transactions on Control Systems Technology*, vol. 25, no. 4, pp. 1257–1263, 2017.

[375] D. A. Shell and M. J. Mataric, "On foraging strategies for large-scale multi-robot systems," *IEEE/RSJ International Conference on Intelligent Robots and Systems*, pp. 2717–2723, 2006.

[376] D. Shevitz and B. Paden, "Lyapunov stability theory of nonsmooth systems," *IEEE Transactions on Automatic Control*, vol. 39, no. 9, pp. 1910–1914, Sep. 1994.

[377] D. Shishika and V. Kumar, "Local-game decomposition for multi-player perimeter-defense problem," *IEEE Conference on Decision and Control*, pp. 2093–2100, 2018.

[378] A. Sih, "Predator-prey space use as an emergent outcome of a behavioral response race," *Ecology of Predator-Prey Interactions*, vol. 11, pp. 240–255, 2005.

[379] N. Sillero, "What does ecological modelling model? A proposed classification of ecological niche models based on their underlying methods," *Ecological Modelling*, vol. 222, no. 8, pp. 1343–1346, 2011.

[380] J. V. Singh and C. J. Lumsden, "Theory and research in organizational ecology," *Annual Review of Sociology*, vol. 16, no. 1, pp. 161–195, 1990.

[381] F. Sistler, "Robotics and intelligent machines in agriculture," *IEEE Journal on Robotics and Automation*, vol. 3, no. 1, pp. 3–6, 1987.

[382] B. Smith, A. Howard, J. McNew, J. Wang, and M. Egerstedt, "Multi-robot deployment and coordination with embedded graph grammars," *Autonomous Robots*, vol. 26, no. 1, pp. 79–98, 2009.

[383] B. Smith, M. Egerstedt, and A. Howard, "Automatic generation of persistent formations for multi-agent networks under range constraints," *Mobile Networks and Applications*, vol. 14, no. 3, pp. 322–335, 2009.

[384] S. L. Smith, M. Schwager, and D. Rus, "Persistent robotic tasks: Monitoring and sweeping in changing environments," *Transactions on Robotics*, vol. 28, no. 2, pp. 410–426, 2012.

[385] T. M. Smith, R. L. Smith, and I. Waters, *Elements of Ecology*. Benjamin Cummings, San Francisco, CA, 2012.

[386] D. Song, N. Qin, Y. Xu, C. Y. Kim, D. Luneau, and K. Goldberg, "System and algorithms for an autonomous observatory assisting the search for the ivory-billed woodpecker," *IEEE International Conference on Automation Science and Engineering*, 2008.

[387] T. R. E. Southwood and P. A. Henderson, *Ecological Methods*. John Wiley & Sons, Hoboken, NJ, 2009.

[388] A. Spears, A. M. Howard, B. Schmidt, M. Meister, M. West, and T. Collins, "Design and development of an under-ice autonomous underwater vehicle for use in polar regions," *IEEE St. John's OCEANS*, pp. 1–6, 2014.

[389] E. Squires, P. Pierpaoli, and M. Egerstedt, "Constructive barrier certificates with applications to fixed-wing aircraft collision avoidance," *EEE Conference on Control Technology and Applications*, pp. 1656–1661, 2018.

[390] S. Squyres, *Roving Mars: Spirit, Opportunity, and the Exploration of the Red Planet*. Hachette Books, New York, NY, 2006.

[391] M. Srinivasan, S. Coogan, and M. Egerstedt, "Control of multi-agent systems with finite time control barrier certificates and temporal logic," *IEEE Conference on Decision and Control*, pp. 1991–1996, 2018.

[392] M. Steinberg, J. Stack, and T. Paluszkiewicz, "Long duration autonomy for maritime systems: Challenges and opportunities," *Autonomous
Robots*, vol. 40, no. 7, pp. 1119–1122, 2016.

[393] S. Still and D. Precup, "An information-theoretic approach to curiosity-
driven reinforcement learning," *Theory in Biosciences*, vol. 131, no. 3,
pp. 139–148, Sep. 2012.

[394] N. E. Stork, "Measuring global biodiversity and its decline," *Biodiversity II: Understanding and Protecting Our Biological Resources*,
vol. 41, pp. 41–68, 1997.

[395] S. Stramigioli, "Energy-aware robotics," in *Mathematical Control Theory I*, Lecture Notes in Control and Information Sciences, edited by
M. K. Camlibel, A. A. Julius, R. Pasumarthy, and J.M.A. Scherpen,
pp. 37–50. Springer, Berlin/Heidelberg, Germany, 2015.

[396] J. Sun, T. Mehta, D. Wooden, M. Powers, J. Regh, T. Balch,
and M. Egerstedt, "Learning from examples in unstructured, outdoor environments," *Journal of Field Robotics*, vol. 23, no. 11–12,
pp. 1019–1036, Nov. 2006.

[397] S. Susca, F. Bullo, and S. Martinez, "Monitoring environmental boundaries with a robotic sensor network," *IEEE Transactions on Control
Systems Technology*, vol. 16, no. 2, pp. 288–296, 2008.

[398] R. S. Sutton and A. G. Barto, *Reinforcement Learning: An Introduction*. MIT Press, Cambridge, MA, 2018.

[399] D. T. Swain, I. D. Couzin, and N. E. Leonard, "Real-time feedback-
controlled robotic fish for behavioral experiments with fish schools,"
Proceedings of the IEEE, vol. 100, no. 1, pp. 150–163, 2011.

[400] K. Szwaykowska and F. Zhang, "A lower bound on navigation error for
marine robots guided by ocean circulation models," *IEEE/RSJ International Conference on Intelligent Robots and Systems*, pp. 3583–3588,
2011.

[401] H. G. Tanner, A. Jadbabaie, and G. J. Pappas, "Flocking in fixed
and switching networks," *IEEE Transactions on Automatic Control*,
vol. 52, no. 5, pp. 863–868, 2007.

[402] A. G. Tansley, "The use and abuse of vegetational concepts and terms,"
Ecology, vol. 16, pp. 204–307, 1935.

[403] J. G. Teer, "Review of conservation biology: The science of scarcity and diversity," *Journal of Wildlife Management*, vol. 52, pp. 570–572, 1988.

[404] R. Tellez, "A thousand robots for each student: Using cloud robot simulations to teach robotics," in *Robotics in Education*, edited by M. Merdan, W. Lepuschitz, G. Koppensteiner, and R. Balogh, pp. 143–155. Springer International Publishing, Cham, Switzerland, 2017.

[405] S. Thrun, W. Burgard, and D. Fox, *Probabilistic Robotics*. MIT Press, Cambridge, MA, 2000.

[406] N. Tinbergen, *Social Behaviour in Animals (Psychology Revivals): With Special Reference to Vertebrates*. Psychology Press, Hove, UK, 2013.

[407] P. Tokekar, E. Branson, H. J. Vander, and V. Isler, "Tracking aquatic invaders: Autonomous robots for monitoring invasive fish," *IEEE Robotics & Automation Magazine*, vol. 20, no. 3, pp. 33–41, 2013.

[408] C. J. Tomlin, G. J. Pappas, and S. S. Sastry, "Conflict resolution for air traffic management: A study in multiagent hybrid systems," *IEEE Transactions on Automatic Control*, vol. 43, no. 4, pp. 509–521, 1998.

[409] A. S. Tonneau, N. Mitton, and J. Vandaele, "How to choose an experimentation platform for wireless sensor networks? A survey on static and mobile wireless sensor network experimentation facilities," *Ad Hoc Networks*, vol. 30, pp. 115–127, 2015.

[410] K. Toussaint, N. Pouliot, and S. Montambault, "Transmission line maintenance robots capable of crossing obstacles: State-of-the-art review and challenges ahead," *Journal of Field Robotics*, vol. 26, no. 5, pp. 477–499, 2009.

[411] L. W. Tu, "Bump functions and partitions of unity," in *An Introduction to Manifolds*, pp. 127–134. Springer, 2008.

[412] H. Tuomisto, "A diversity of beta diversities: Straightening up a concept gone awry. Part 1. Defining beta diversity as a function of alpha and gamma diversity," *Ecography*, vol. 33, no. 1, pp. 2–22, 2010.

[413] M. G. Turner, "Landscape ecology: The effect of pattern on process," *Annual Review of Ecology and Systematics*, vol. 20, no. 1, pp. 171–197, 1989.

[414] P. Twu, P. Martin, and M. Egerstedt, "Graph process specifications for hybrid networked systems," *Journal of Discrete Event Dynamic Systems*, vol. 22, no. 4, pp. 541–577, Dec. 2012.

[415] P. Twu, Y. Mostofi, and M. Egerstedt, "A measure of heterogeneity in multi-agent systems," *American Control Conference*, pp. 3972–3977, 2014.

[416] E. Ugur, M. R. Dogar, M. Cakmak, and E. Sahin, "Curiosity-driven learning of traversability affordance on a mobile robot," *IEEE International Conference on Development and Learning*, Oct. 2007.

[417] J. van den Berg, M. Lin, and D. Manocha, "Reciprocal velocity obstacles for real-time multi-agent navigation," *IEEE International Conference on Robotics and Automation*, pp. 1928–1935, 2008.

[418] V. Vladimerou, A. Stubbs, J. Rubel, A. Fulford, and G. Dullerud, "Multivehicle systems control over networks," *IEEE Control Systems Magazine*, vol. 26, no. 3, pp. 56–69, 2006.

[419] J. Waage and D. Greathead, "Insect parasitoids," *Symposium of the Royal Entomological Society of London*, 1986.

[420] G. Wagner and H. Choset, "M*: A complete multirobot path planning algorithm with performance bounds," *IEEE/RSJ International Conference on Intelligent Robots and Systems*, pp. 3260–3267, 2011.

[421] M. Wall, "*Opportunity* rover still silent on Mars, 4 months after epic dust storm began," *Space.com*, Oct. 2018.

[422] W. G. Walter, *The Living Brain*. W. W. Norton, New York, NY, 1953.

[423] L. Wang, A. D. Ames, and M. Egerstedt, "Multi-objective compositions for collision-free connectivity maintenance in teams of mobile robots," *IEEE Conference on Decision and Control*, pp. 2659–2664, 2016.

[424] L. Wang, A. D. Ames, and M. Egerstedt, "Safety barrier certificates for collision-free multirobot systems," *IEEE Transactions on Robotics*, vol. 33, no. 3, pp. 661–674, 2017.

[425] L. Wang, A. D. Ames, and M. Egerstedt, "Safety barrier certificates for heterogeneous multi-robot systems," *American Control Conference*, pp. 5213–5218, Jul. 2016.

[426] L. Wang, E. Theodorou, and M. Egerstedt, "Safe learning of quadrotor dynamics using barrier certificates," *IEEE International Conference on Robotics and Automation*, May 2018.

[427] T. G. Whitham and S. Mopper, "Chronic herbivory: Impacts on architecture and sex expression of pinyon pine," *Science*, vol. 228, no. 4703, pp. 1089–1091, 1985.

[428] N. L. Wiggins, C. McArthur, and N. W. Davies, "Diet switching in a generalist mammalian folivore: Fundamental to maximising intake," *Oecologia*, vol. 147, no. 4, pp. 650–657, 2006.

[429] E. O. Wilson, *The Diversity of Life*. W. W. Norton & Company, New York, NY, 1999.

[430] E. O. Wilson, *The Future of Life*. Vintage, New York, NY, 2002.

[431] S. Wilson, P. Glotfelter, L. Wang, S. Mayya, G. Notomista, M. Mote, and M. Egerstedt, "The Robotarium: Globally impactful opportunities, challenges, and lessons learned in remote-access, distributed control of multirobot systems," *IEEE Control Systems Magazine*, vol. 40, no. 1, pp. 26–44, 2020.

[432] M. Wisse, G. Keliksdal, J. Van Frankenhyyzen, and B. Moyer, "Passive-based walking robot," *IEEE Robotics & Automation Magazine*, vol. 14, no. 2, pp. 52–62, 2007.

[433] B. Wittwer and M. A. Elgar, "Cryptic castes, social context and colony defense in a social bee, *Tetragonula carbonaria*," *Ethology*, vol. 124, no. 8, pp. 617–622, 2018.

[434] D. Wooden, M. Powers, M. Egerstedt, H. I. Christensen, and T. Balch, "A modular, hybrid system architecture for autonomous, urban driving," *Journal of Aerospace Computing, Information, and Communication*, vol. 4, no. 12, pp. 1047–1058, 2007.

[435] M. Worall, A. J. Jamieson, A. Holford, R. D. Neilson, M. Player, and P. Bagley, "A variable buoyancy system for deep ocean vehicles," *IEEE St. John's OCEANS*, pp. 1–6, 2007.

[436] X. Xu, P. Tabuada, J. W. Grizzle, and A. D. Ames, "Robustness of control barrier functions for safety critical control," *IFAC World Congress*, vol. 48, no. 27, pp. 54–61, 2015.

[437] P. Yang, R. A. Freeman, G. J. Gordon, K. M. Lynch, S. Srinivasa, and R. Sukthankar, "Decentralized estimation and control of graph

connectivity for mobile sensor networks," *Automatica*, vol. 46, no. 2, pp. 390–396, 2010.

[438] C. Yu, B. D. O. Anderson, S. Dasgupta, and B. Fidan, "Control of minimally persistent formations in the plane," *SIAM Journal on Control and Optimization*, vol. 48, no. 1, pp. 206–233, 2009.

[439] M. Zavlanos, M. Egerstedt, and G. J. Pappas, "Graph theoretic connectivity control of mobile robot networks," *Proceedings of the IEEE*, vol. 99, no. 9, pp. 1525–1540, 2011.

[440] F. Zhang and N. E. Leonard, "Coordinated patterns of unit speed particles on a closed curve," *Systems and Control Letters*, vol. 56, no. 6, pp. 397–407, 2007.

Index